T0305169

Russian Oil Companies in an Evolving World

Russian Oil Companies in an Evolving World
The Challenge of Change

Indra Overland

Research Professor and Head of the Centre for Energy Research at the Norwegian Institute of International Affairs (NUPI), Oslo, Norway

Nina Poussenkova

Senior Researcher, Primakov Institute of World Economy and International Relations (IMEMO), Russian Academy of Sciences and Researcher, ENERPO Research Centre, European University in Saint Petersburg, Russia

Edward Elgar
PUBLISHING

Cheltenham, UK • Northampton, MA, USA

Published by
Edward Elgar Publishing Limited
The Lypiatts
15 Lansdown Road
Cheltenham
Glos GL50 2JA
UK

Edward Elgar Publishing, Inc.
William Pratt House
9 Dewey Court
Northampton
Massachusetts 01060
USA

A catalogue record for this book
is available from the British Library

Library of Congress Control Number: 2020940506

This book is available electronically in the **Elgar**online
Social and Political Science subject collection
http://dx.doi.org/10.4337/9781788978019

ISBN 978 1 78897 800 2 (cased)
ISBN 978 1 78897 801 9 (eBook)

Printed and bound by CPI Group (UK) Ltd, Croydon, CR0 4YY

Contents

Figures

Tables

About the authors

Indra Overland is Research Professor and Head of the Centre for Energy Research at the Norwegian Institute of International Affairs (NUPI). He was previously Head of the Russia, Eurasia and Arctic Research Group at NUPI. Overland completed his PhD at the University of Cambridge and has published extensively on post-Soviet energy issues. He is the co-editor of *Caspian Energy Politics* (Routledge 2010) and *International Arctic Petroleum Cooperation: Barents Sea Scenarios* (Routledge 2015) and co-author of *Financial Sanctions Impact Russian Oil* (Oil & Gas Journal 2015) and *Ranking Oil, Gas and Mining Companies on Indigenous Rights in the Arctic* (Árran 2016).

Nina Poussenkova is Senior Researcher at the Institute of World Economy and International Relations (IMEMO), Russian Academy of Sciences, where she heads the IMEMO Oil and Gas Dialogue Forum, and researcher at the ENERPO Centre of the European University in Saint Petersburg. In the 1990s, she worked at the Centre for Foreign Investment and Privatization and at the investment banks Salomon Brothers and Lazard Freres. From 2006 to 2008, she served as Director of the Energy Programme of the Carnegie Moscow Centre. Poussenkova has also led WWF-Russia projects on Russian companies and corporate citizenship in the twenty-first century and on benchmarking of Russian refineries. She previously taught at the Gubkin State University of Oil and Gas. She is the author of *Russian Oil Companies* (Encyclopaedia of Mineral and Energy Policy, Springer 2016), *The Rosneftization of the Russian Oil Sector* (Oxford Energy Forum 2016) and the co-author of *Arctic Petroleum: Local CSR Perceptions in the Nenets Region of Russia* (Social Responsibility Journal 2017) and *Russia: Public Debate and the Petroleum Sector* (Public Brainpower, Palgrave 2017).

Acknowledgements

We would like to thank the following people for their advice and input on various elements included in the book: Saule Aripova, Angel Barajas, Gulzhan Begeyeva, Helge Blakkisrud, Tatyana Dzhiganshina, Daniel Fjaertoft, Vladimir Gelman, Jakub Godzimirski, Dmitry Goncharov, James Henderson, Talgat Ilimbek uulu, Javlon Juraev, Galina Khegay, Daniyar Kussainov, Mikhail Krutikhin, Julia Loe, Arild Moe, Alesia Prachakova, Haakon Fossum Sagbakken, Abdyrakhman Sulaimanov, Emma Wilson and language editors and anonymous reviewers from Scribendi and Edward Elgar. All translations are by the authors unless stated otherwise. This book is a product of the RusChange project, which is financed by the PETROSAM programme of the Research Council of Norway. The introductory, concluding and Gazprom Neft chapters were funded by the project 'Is this Russia's Kodak Moment?', which is financed by the NORRUSS programme of the Research Council of Norway.

Terminology and acronyms

ADR	American depository receipt (vehicle for trading non-American stocks on American stock exchanges)
AGM	annual general meeting (gathering of all shareholders)
APG	associated petroleum gas (gas found along with oil)
ARCO	Atlantic Richfield Company
BCG	Boston Consulting Group
billion	a thousand million
BP	(formerly) British Petroleum
CEO	chief executive officer
CIS	Commonwealth of Independent States
CNPC	China National Petroleum Corporation
CSR	corporate social responsibility
E&P	exploration and production
EITI	Extractive Industries Transparency Initiative
EOL	Essar Oil Limited
ESPO	Eastern Siberia–Pacific Ocean pipeline
EU	European Union
FDI	foreign direct investment
GAAP	generally accepted accounting principles
GDP	gross domestic product
GHG	greenhouse gas
GRI	Global Reporting Initiative
HSE	health, safety and environment
IAS	international accounting standards
IEA	International Energy Agency
IFRS	international financial reporting standards
IOC	international oil company

IPO	initial public offering
IRENA	International Renewable Energy Agency
IT	information technology
JSC	joint stock company
KNOC	Korean National Oil Corporation
LNG	liquefied natural gas
LSE	London Stock Exchange
M&A	mergers and acquisitions
MD&A	management discussion and analysis
MoU	memorandum of understanding
MRH	Mineralol Rohstoff Handel
NFK	Neftyanaya Finansovaya Kompaniya
NIOC	National Iranian Oil Company
NIS	Naftna Industrija Srbije
NOC	national oil company
NYSE	New York Stock Exchange
OECD	Organisation for Economic Co-operation and Development
OJSC	open joint stock company
ONACO	Orenburgskaya Neftyanaya Kompaniya (Orenburg Oil Company)
ONGC	Oil and Natural Gas Corporation (India)
OPEC	Organization of Petroleum Exporting Countries
PDVSA	Petróleos de Venezuela
PSA	production-sharing agreement
R&D	research and development
Rosnedra	Federalnoe agenstvo po nedropolzovaniyu (Federal Agency for Subsoil Use)
R/P	reserves to production ratio
RTS	Russian Trading System
RUB	Russian roubles
S&P	Standard and Poor's Financial Services
SEC	Securities and Exchange Commission

SIDANCO	Sibirskaya-dalnevostochnaya Neftyanaya Kompaniya (Siberian-Far Eastern Oil Company)
TNK	Tyumenskaya Neftyanaya Kompaniya (Tyumen Oil Company)
TRN	Total Refinery Netherlands
TSR	total shareholder return
UN	United Nations
UNESCO	United Nations Educational, Scientific and Cultural Organization
USSR	Union of Soviet Socialist Republics
WWF	World Wildlife Fund
YUNKO	Yuzhnaya Neftyanaya Kompaniya (Southern Oil Company)

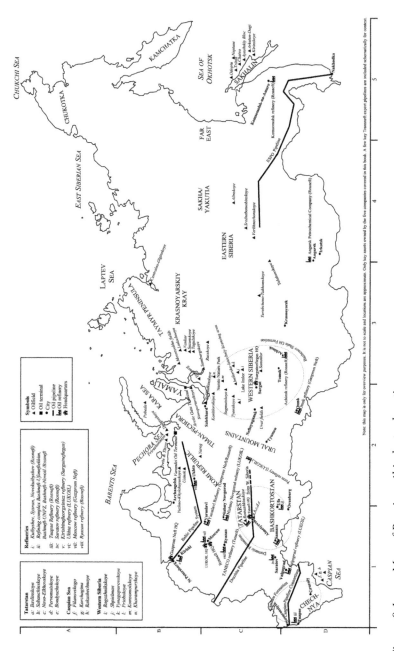

Figure 0.1   Map of Russian oil industry locations

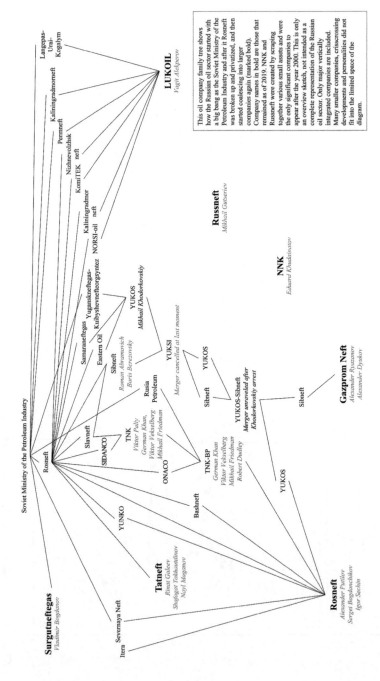

Figure 0.2   Russian oil company family tree

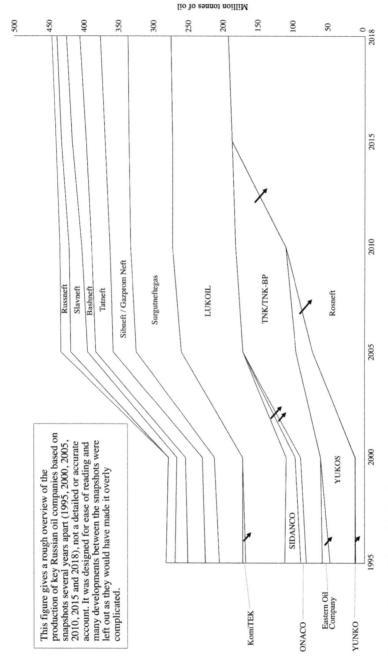

This figure gives a rough overview of the production of key Russian oil companies based on snapshots several years apart (1995, 2000, 2005, 2010, 2015 and 2018), not a detailed or accurate account. It was designed for ease of reading and many developments between the snapshots were left out as they would have made it overly complicated.

Million tonnes of oil

Figure 0.3   Oil production by company

1. Introduction: can Russia and its oil companies handle change?

This book examines Russia's capacity to respond to a changing world, as seen through the lens of the country's oil companies. The reasons for this choice of topic are the central role of the petroleum sector in the Russian economy, the great importance of Russian oil and gas to the world's energy supply and the rapid pace of change in the global energy industry. The petroleum sector accounts for 40% of the Russian state's income (TASS 2018) and employs over 1.1 million people, with more than 400 000 working for Gazprom alone. Russia is also the world's largest energy exporter, and Russian foreign policy is interwoven with market access, cooperation and investments in the energy sector (Casier 2011; Hendrix 2015; Jirušek et al. 2017; Kalehsar and Telli 2017; Proedrou 2017; Wigell and Vihma 2016).

The Kremlin maintains a firm grip on Russia's oil and gas industry by controlling the largest producers, Gazprom and Rosneft, by setting framework conditions for the operations of other companies and by doling out tax breaks and other privileges to preferred actors. However, the actual work in the petroleum sector is done by the companies themselves, regardless of whether they are wholly or partially privately owned. How these companies cope with a changing world is decisive for the fate of the Russian petroleum industry and thus for Russia. This book, therefore, focuses on these organizations rather than the role of the state.

THE CHALLENGE OF CHANGE

Russian oil companies must deal with many types of change, such as oil price swings, currency fluctuations, the rise of shale oil, Western sanctions over the conflict in Ukraine, the shift of oil consumption growth to Asia and the increasing salience of climate policy. In the long term, the most important of these may be climate policy, which is upending the global energy system.

World energy markets are subject to continuous incremental change and occasional technology-driven tsunamis referred to as 'energy transitions' (Fouquet 2016; Fouquet and Pearson 2012; Grubler 2012; Meadowcroft 2009; Smil 2010). Past energy transitions have been associated with the emergence of disruptive technologies, such as electricity, the internal combustion engine and nuclear power. Climate policy is now driving a new energy transition towards renewable energy, especially solar and wind power. This transition may change international demand for the products of Russian oil companies, in addition to putting pressure on the companies to reduce their direct emissions. How Russia and its oil companies deal with this and other changes will be one determinant of the country's strength in global affairs (Overland and Kjaernet 2009).

Many East Asian and Western companies are locked in battles of innovation and counter-innovation – for example Apple, Huawei and Samsung in telephony; or Build Your Dreams (BYD), Nissan and Tesla in road transport. To play a role in the evolving world, Russia's major companies must be able to manage change and, ideally, even become its drivers. If there is one area where they have a chance of doing this, it is in the petroleum sector, where they have such a strong position and rich history. It dates all the way back to the first oil well and refinery in the town of Ukhta in the Komi Republic in 1745. Russia's current position as the world's biggest combined oil and gas exporter could also, in principle, give it an upper hand in driving change (Poussenkova and Overland 2018, p. 261).

HYPOTHESIS: RUSSIAN OIL COMPANIES LACK FORESIGHT

This book seeks to answer two main questions. First: How are Russian oil companies tackling the changing global context? This question concerns both how Russian actors think about the future and plan for it, and how they handle change once it happens. In other words, this question is about the adaptability of Russian actors. Second: How are the companies themselves changing? This concerns the people in charge, their corporate culture and political connections and the companies' oil reserves.

Our working hypothesis is that Russian organizations, including oil companies, are not good at foreseeing and adapting to change. Gustafson (2012, p. 2) writes that 'despite two decades of tumultuous changes, the pull of the past – of the assets and mindsets of the Soviet legacy – remains strong. Yet the Russian oil industry is now exposed to a global energy system that is itself experiencing a revolution. Consequently, it too is under pressure to change.' Dixon (2008, p. 42) adds that 'in transition economies, many of the big state-owned enterprises have become corporate dinosaurs, entrenched in old behaviours and unable to make the first steps of change to adapt to a changing environment' (also see Peng 2000).

We present our hypothesis in the spirit of Popper – as something to try to shoot down. However, it is not difficult to find anecdotal evidence from Russian history to back it up. The Communist Party appeared unprepared for the steep decline of the oil price in the 1980s, and both the party and Soviet society at large seemed to be caught unawares by the unravelling of the Soviet Union in 1991, to which the low oil price greatly contributed (Friedman 2006; Kotkin 2008; Reynolds and Kolodziej 2008). Since then, Russian oil companies have matured, worked closely with their international counterparts and expanded into foreign upstream and downstream markets (Henderson and Ferguson 2014). Nonetheless, for years, they continued to deny the significance of the shale revolution (Davydova 2017b; Grealy 2012; Spencer and Hansen 2012). Russian energy actors (along with some of their Western colleagues) have also been among the world's most entrenched climate sceptics, even though Russia is a signatory to both the Kyoto Protocol and the Paris Agreement.

One possible explanation for this laggardness could be the age and gender composition of the top managers of the Russian oil companies. They tend to be exclusively elderly men, few of whom have any foreign education (Figures 1.1–1.3).

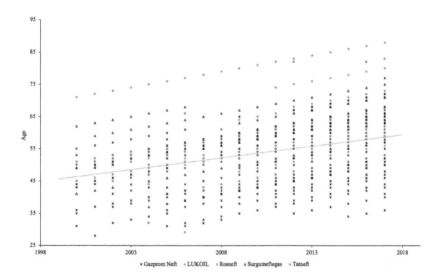

Figure 1.1   Age of CEOs and board members of Russian oil companies

Source: Compiled by the authors based on a large number of sources.

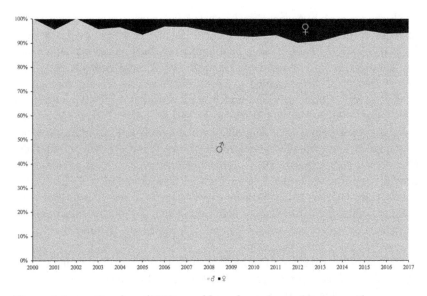

*Figure 1.2   Gender of CEOs and board members of Russian oil
               companies*

Source: Compiled by the authors based on a large number of sources.

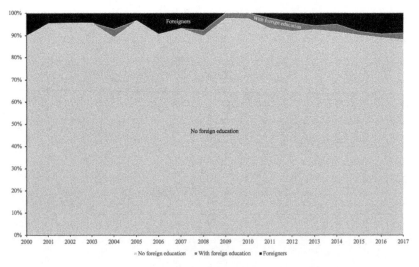

*Figure 1.3   Foreign education among CEOs and board members of
               Russian oil companies*

Source: Compiled by the authors based on a large number of sources.

Another possible explanation for the laggardness noted above could be that Russian business is characterized by a grab-and-run mentality and, therefore, simply does not care much about long-term change. This impression is supported by the rough experiences of some foreign companies and businesspeople in Russia (Bogdanova 2015; Dahlgren 2009; Fabry and Zeghni 2002; Liuhto 2010). The post-Soviet period witnessed upheaval and socio-economic instability; property rights, in particular, were insecure and susceptible to revision (Adachi 2010; Gans-Morse 2017; Ledeneva 2013; Person 2016; Walker 2015). Fortunes were amassed and lost in short shrift, and investment horizons were correspondingly limited. The lifestyles of many Russian businesspeople seemed to indicate that they considered it more important to enjoy the moment than to secure their wealth, as they found it difficult to control or predict what would happen in the long term anyway. If this is a correct reading of post-Soviet society, Russian oil companies should, indeed, be expected to be more concerned with quick profits than with anticipating and preparing for long-term global change.

As a corollary to the assumption of post-Soviet short-termism, it could be argued that Russian oil companies are geared towards milking what remains of the Soviet resources and infrastructure rather than doing the demanding work needed to create something new and financially sustainable. Few discoveries of major oilfields have been made in the country in the past 30 years; instead, the oil companies have subsisted on the discoveries from the Soviet period and legacy fields, such as Samotlor and Vankor. Also the infrastructure was largely built during socialism, in particular the oil and gas pipelines. At times, Russia's entire post-Soviet society has seemed to be one large asset-stripping operation, selling off everything from scrap metal to human resources – and oil. In this regard, Russia is a rentier state in a very literal sense: not only does it live off the easy-come, easy-go income from oil, but within the petroleum sector it is milking the infrastructure, competencies and structures inherited from the *ancien régime*. According to Gustafson (2012, p. 5),

> Russia's oil industry and the Russian state are not well-prepared to deal with the coming challenge. They have spent the last two decades competing for control of the inherited oil assets and rents instead of cooperating to modernize the industry better and prepare for the next stage ahead . . . The result is an industry that compared with its world peers, lags behind the rapidly moving front of a global oil business that is in the midst of a technological and managerial revolution.

Another factor that may arguably weaken the capacity of Russian actors to deal with change is conservatism. The longevity of the Soviet system, and the tsarist system before it, could be interpreted as an indication of a deep conservative current in Russian society. Marxist communism originated as a fashionable ideology in the West but later metamorphosed into a range of socialist and social-democratic ideologies. When this fashion reached China, Cuba, Russia and Vietnam, it instead became entrenched and served as the static organizing principle for the state and society for over half a century.

It seems that many Russians like things to stay the way they are and, therefore, tend to stick their heads in the sand rather than face changing markets, technologies and social and economic systems (Kennaway 2000). It is difficult to confirm or reject such an argument unequivocally. However, through a systematic review of empirical data, we can explore the argument in question in greater detail.

SUDDEN AND GRADUAL CHANGE

We can distinguish between two main types of change faced by oil companies. The first type is sudden, taking from a few days to a few years and relating to dramatic political and economic events, such as wars, coups, international conflicts, oil price spikes and stock market crashes. Notable examples include the 2008 financial crisis, the 2014 oil price collapse, the conflict between Russia and the West over Ukraine and the ensuing sanctions targeting the Russian energy sector.

The second type of change takes place over a longer period – years or even decades. Such change may be gradual and unnoticeable, or it may be the sum of many small increments, sometimes in a pattern of two steps forward and one step back. Although it is less dramatic than a sudden change, its consequences can be as great, and the moment of realization of what is happening can be sudden. Notable examples of gradual change in the petroleum sector include the expansion of offshore oil and gas extraction in the OECD (Organisation for Economic Co-operation and Development) countries during the 1970s (Correljé 2018; Overland 2018; Wright and Boué 2018), declining production from the Soviet legacy oilfields in Western Siberia (Gustafson 2012), the shift in oil demand growth from the West to East Asia (Overland 2015), the rise of shale oil (Boersma and Johnson 2012), the increasing interconnectedness of regional gas markets due to the expansion of liquefied natural gas (LNG) (Wright 2017) and the incipient electrification of transport (Sovacool 2017).

Sudden changes are inevitably linked to gradual developments, and gradual developments may take new directions due to sudden changes. It can, therefore, be difficult to classify events such as the Paris Agreement as sudden or gradual. After years of failed negotiations in Bali, Cancun, Copenhagen, Doha and Warsaw, it was difficult to predict whether an agreement would be reached in Paris in 2015 – until it actually happened (Hufbauer and Kim 2010). On the other hand, the push towards a more comprehensive and harder-hitting climate policy has been consistent since the early 1990s (Campbell 2013; Sprinz et al.

2016). One could, therefore, argue that it was likely that an agreement would be signed sooner or later after years of glacial negotiations.

ANALYTICAL TOPICS

In each of the company chapters that make up the bulk of this volume, we try to touch on the same topics. They can be divided into two groups. Group I topics profile each company in terms of its role in Arctic and offshore oil extraction, internationalization, transparency and innovation. Group II topics have been selected specifically to assess the companies' adaptability to a changing global environment. They include some of the major international energy developments that have affected Russian oil companies since the turn of the millennium and their responses to these developments, including the shale revolution, oil price volatility, sanctions and climate policy.

Group I Analytical Topics

The Arctic
With the decline of the onshore Soviet legacy fields, Russian companies are being forced to migrate northwards and offshore. From 2016 to 2017, Russia's Arctic oil production grew by 10% (TASS 2017). President Putin (cited in Kramer 2011) stated that the Kara Sea alone would require USD 500 billion in investments. The Arctic is financially and technologically challenging for Russian companies on their own, and international oil companies have had a standing invitation to help develop Russia's Arctic petroleum frontier (Overland et al. 2013). Consequently, the Arctic is perhaps the most important arena for interaction between Russian companies and their international peers. Despite this, the sanctions and falling oil prices largely undermined the potential for interaction from 2014 onwards (Aalto 2016).

Offshore
Russian petroleum development is increasingly taking place offshore, namely in the Arctic Ocean, the Caspian Sea and the Pacific Ocean. Not coincidentally, the sanctions against Russia over its role in the conflict in Ukraine specifically target Arctic and offshore oil and gas developments – an attempt to hit where it hurts (Fjaertoft and Overland 2015).

An important feature of offshore developments is that their costliness renders them vulnerable to oil price volatility (Overland et al. 2015). This is doubly true when they are located in the Arctic or other remote areas with harsh climatic conditions, such as the Sea of Okhotsk. The Shtokman gas and condensate field in the Barents Sea once received considerable attention from both Russian and foreign companies but was shelved because of the costs of developing the field as well as falling gas prices (Henderson and Moe 2016; Overland et al. 2015). When oil and gas prices drop, the competition between petroleum provinces intensifies, and the most expensive areas see a decline in investment. Russian Arctic offshore resources are especially exposed because of compounded risk. Like the rest of the Arctic, costs are driven up by the harsh climate and distance to markets; in Russia, political risk comes on top of this.

Internationalization
For Russian companies, the Arctic and offshore are frontiers where they can benefit from partnering with international companies (Aalto 2016). Some deals between the Russian and international companies are in the range of tens of billions of dollars, such as the alliance between ExxonMobil and Rosneft forged in 2011 or the agreement between Gazprom and China National Petroleum Corporation (CNPC) on gas exports signed in 2014 (Lunden et al. 2013; Sharples 2016). These relationships serve as an important interface between Russia and the world. They have sometimes been characterized by conflict and trouble, as in the relationship between BP and Gazprom over the Kovykta gas field in Siberia or between Shell and Gazprom in the areas off Sakhalin Island (Henderson and Moe 2016; Kyj and Kyj 2010; Locatelli 2006; Sevastyanov 2008). Partly because of such past quarrels and partly because of the Western sanctions against Russia, there has been a shift away from part-nerships with Western companies to Chinese and other Asian oil companies (Overland and Kubayeva 2018).

The other side of this interface is the Russian oil companies' own projects overseas. For several companies, such projects have been important to diffuse risk and gain experience, access new petroleum reserves, strengthen political ties with strategic partner countries such as Venezuela or as a token of a desire to be seen as serious actors on the international scene. As a result, the Russian oil companies are present in many parts of the world (Figure 1.4).

Transparency
The more involved Russian oil companies become with Western companies and institutions, the more important it is for the Westerners that the Russian companies are seen as transparent. This is especially true when the Russian companies are listed on foreign stock exchanges with many international investors acquiring stakes. Table 1.1 gives Ivolga et al.'s (2018) assessment of the transparency of Russian companies in the petroleum sector, including small and large ones as well as subsidiaries. In this book, we provide an alter-native perspective on the major companies italicized in the table.

Number of Companies

1   2   3

Figure 1.4   *Russian petroleum foreign direct investment (FDI) projects around the world (2018)*

Note: Light grey = one Russian oil company present; dark grey = two; black = three.
Source: Compiled by the authors based on a large number of sources.

Table 1.1   *Transparency ratings of Russian oil and gas companies*

Company	Score	Company	Score
Zarubezhneft	5.3	Slavneft	3.5
Novatek	5.1	Transneft	2.5
RussNeft	5.1	Neftisa	1.4
Bashneft	5.1	Salym Petroleum	1.0
Rosneft	*4.9*	Transoil	0.9
Tatneft	*4.6*	Tomskneft	0.0
Gazprom	4.4	LUKOIL Garant	0.0
Surgutneftegas	*4.3*	Transnafta	0.0
LUKOIL	*4.1*	Eurasia Drilling	0.0
Gazprom Bureniye	3.7		

Note: Score 0–10, where 10 denotes most transparent. The companies covered in this book have been italicized.
Source: Ivolga et al. (2018).

Innovation

Innovation is closely linked to foresight and thus serves as a bridge to Group II topics. In order to innovate, companies must see and understand how energy demand may develop in the future. President Putin has repeatedly called on major Russian companies to be more innovative (Putin 2017b), and Dmitry Medvedev made innovation the centrepiece of his presidency (Overland 2011). Nonetheless, there is a lingering impression that Russian oil companies are not innovative. It therefore makes sense for us to pay attention to any innovation efforts on the part of the companies.

Group II Analytical Topics

The purpose of the following set of analytical topics is to assess the adaptability of Russian oil companies to a changing global environment, in particular their responses to major international energy developments after the turn of the millennium.

Shale oil and gas

The North American shale revolution shook international markets (Figure 1.5). This development was particularly important for Russia, both as the world's largest combined oil and gas exporter and as a country that opposes the United

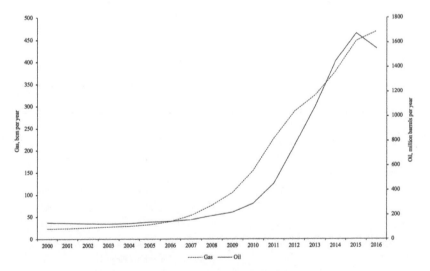

Figure 1.5   *Growth in US shale oil and gas production*

Source: EIA (2019a, 2019b); IEA (2019).

States on many fronts (Overland et al. 2013, p. 146). However, for a long time, prominent Russian actors remained outspokenly sceptical about shale oil and gas (Davydova 2017b; Grealy 2012; Spencer and Hansen 2012). Deputy Head of Gazprom Alexander Medvedev referred to shale gas as 'a bubble', and chief executive officer (CEO) Alexey Miller argued that shale gas would remain a luxurious side dish: 'If you like *foie gras*, that doesn't mean you no longer need a regular steak' (cited in Elder 2012). Shale oil and gas are, therefore, of particular interest in a discussion of how Russian companies relate to the changing world.

Oil price

Oil companies must always be prepared for price fluctuations. From 1999 to 2008, the oil price skyrocketed from USD 20 to USD 140 per barrel; it subsequently fell to USD 40 in 2009 but bounced back again to USD 120 in 2011 and fell again to USD 30 in 2016 (Figure 1.6). As oil projects have long lead times, with decades between an investment decision and decommissioning, fluctuating prices pose a major challenge (Arezki et al. 2017; Clo 2000; Shaukat Khan et al. 2016). Gustafson (2012, p. 4) writes about the long period of rising prices from 1999 onwards: 'For the Russian elites, these were heady times – high oil prices were "here to stay" – and they became cocksure and complacent.' The same could be said of many oil-driven elites around the world.

Figure 1.6   Oil and gas prices

Source: Index Mundi (2020); EIA (2020).

Sanctions
The Western sanctions against Russia over the conflict in Ukraine targeted offshore, deep-water, unconventional and Arctic oil as well as specific Russian oil companies and individuals in the Russian petroleum sector (Fjaertoft and Overland 2015, p. 66).

The conflict in Ukraine and the introduction of sanctions against Russia coincided with the collapse of the oil price in 2014. In the preceding years, Russian oil companies had been incurring debt based on high oil price valuations of their assets and future income. The sanctions thus came at an inopportune moment for the Russian companies, revealing some of the political risks *they* face concerning Russian-Western relations.

Climate policy and energy transition
The most profound and long-term changes underway in the energy sector are driven by climate policy, which is increasing the pressure for a transformation of the energy mix. By 2050, there may be between 40% and 65% renewable energy in the global energy mix, up from 13% in 2012, depending on whether one uses the scenarios of the International Energy Agency (IEA) or the International Renewable Energy Agency (IRENA) (O'Sullivan et al. 2017, p. 9). However, the scenarios of both these organizations envisage the rapid expansion of renewable energy and a corresponding lesser role for fossil fuels. While rising energy consumption may ensure that there is still a significant market for oil and gas, energy efficiency will pull in the other direction, rendering the prospects for oil companies uncertain.

Worldwide, oil companies have been slow to recognize these developments (Ben-Amar and McIlkenny 2015; Besio and Pronzini 2014; Hiatt et al. 2015; Mitchell and Mitchell 2014; Schlichting 2013). ExxonMobil's denial of climate change has received particular attention (Supran and Oreskes 2017). However, some international oil companies – such as Shell, Equinor and Total – have started positioning themselves for an energy transition by beginning to shift some capital from the petroleum sector to solar and wind power. How, then, are Russia's oil companies coping with the implications of climate change and evolving climate policy? Are they experiencing a 'Kodak moment', underestimating the potential of emerging politics and technologies and overestimating future demand for their products (Griffin et al. 2015; van der Ploeg 2016)?

On the one hand, if the companies take their cue from the official policy of the Russian state, they might not be entirely unprepared for the effects of climate policy on energy demand. Compared to China or the United States, for instance, the Russian state has been relatively consistent in its support for international climate policy in the international diplomatic arena. Unlike those two countries, Russia ratified the Kyoto Protocol, thereby enabling it to come into force and saving it (Henry and Sundstrom 2007, p. 47). Russia subsequently over-fulfilled its Kyoto Protocol emissions reduction targets, thus helping compensate for countries that did not fulfil their own targets (Putin 2017a). In 2013, and again in 2015, the Russian government established support schemes for renewable energy (Boute 2016). In 2017, the Working Group on Climate Change and Sustainable Development under the Presidential Administration launched an all-Russian climate week with 422 events across the country (Valeeva 2017). Of direct relevance to this book is the government's signal that it will introduce legislation limiting greenhouse gas (GHG) emissions from Russian companies and establish targets for the use of associated petroleum gas (APG) from oil production (Davydova 2017a).

However, in many other ways, Russia has remained a laggard on climate change and its oil companies might also be expected to perform below average. Russia was one of the last major greenhouse gas emitters to ratify the Paris Agreement, and Russian emissions targets do not match the country's commitments under international agreements (Sharmina 2017).

Like their American counterparts, some prominent Russians have publicly been deeply sceptical about climate change (Tynkkynen and Tynkkynen 2018; Skryzhevska et al. 2015). After visiting Franz Josef Land between the Arctic, Barents and Kara Seas, President Putin declared that climate change had nothing to do with human activity (cited in Farand 2017; Meredith and Cutmore 2017). The Executive Chairman of Rosneft, Igor Sechin, has also publicly expressed scepticism about climate change; arguing that the effect of anthropogenic greenhouse gas emissions cannot compare to those of volcano eruptions or rotting algae and stating that climate change is largely due to 30-million-year natural climatic cycles (Sechin, cited in Armitage 2015). Similar to American President Donald Trump's statement that climate change is a Chinese hoax, some major Russian media have cast climate change as a foreign plot to undermine Russian energy exports or as an American weapon aimed at Russia (Davydova 2017b). Such statements have caused some commentators to become highly critical of Russia, arguing that it is failing to adapt to the new realities of global climate policy: 'In the new geopolitics of renewable energy, post-fossil Russia does not have a value proposition . . . Oil addiction is hard to cure, and Russia is not even trying' (Kraemer 2017).

As this book is about change, the forecasts of Russian governmental energy institutions are particularly relevant. Russian Minister of Energy Alexander Novak stated that electric vehicles will make up only 1% of all cars in the world by 2035 and, therefore, will not have much impact on oil demand (Novak 2016). In 2013, several top Russian energy experts published a global energy forecast that mentions renewables only once and does not mention climate change at all, although the report is 110 pages long (Makarov et al. 2013). The 2014 175-page issue of the same report mentions renewables 18 times and climate change five times (Makarov et al. 2014). Only in the 2016 issue did the report start paying serious attention to these topics, with 24 mentions of renewable energy and 15 mentions of climate change – still not much compared to the attention these topics were attracting in the energy analyses and forecasting of many other countries (Makarov et al. 2016). In the 2019 issue, the mentions of the renewable energy and climate were still at the same level, with renewable energy mentioned 26 times and climate change 15 times (Makarov et al. 2019).

In sum, the Russian state has been sending mixed signals to the country's oil companies. How are they responding? Figure 1.7 gives a first impression of the attention of the companies to climate change. In the chapters dedicated to each company, we look more closely at how they have been handling climate policy and its implications for energy demand.

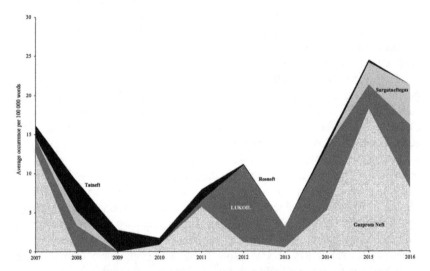

Figure 1.7   *Occurrence of 'climate change OR renewable energy' per 100 000 words in Russian oil and gas company reports (2007–16)*

Source: Compiled by the authors based on a large number of sources.

While we aim to cover all the topics outlined above in every chapter dedicated to a company, each company has a unique history and character, and each chapter, therefore, has a different shape. In addition to the topics outlined here, we touch on several other topics related to the Russian petroleum sector without attempting to provide detailed accounts, as these have already been provided elsewhere in the literature. This includes FDI in the petroleum sector

(Bayulgen 2014), assessments of oil and gas fields and their decline (Grace 2005), pipeline politics (Barysch 2008; Stulberg 2012), the role of energy in foreign policy (Hill 2004), the oligarchs and elites (Balmaceda 2008; Bulavka and Buzgalin 2016; de Graaff 2012; Maury and Liljeblom 2009; Rivera and Rivera 2014), Vladimir Putin (Appel 2008; Balzer 2005; Goldman 2010), economic reform (Gaddy 2013), Dutch disease and the resource curse (Bradshaw 2006; Gaddy and Ickes 2019; Mironov and Petronevich 2015; Tompson 2005) and corruption (Cheloukhine and King 2007; Obydenkova and Libman 2015; Rutland 2015; Smith and Thomas 2015).

SELECTION OF COMPANIES

This book covers the following companies, with one chapter dedicated to each: Rosneft, LUKOIL, Gazprom Neft, Surgutneftegas and Tatneft. These five were chosen because they are the largest Russian oil companies. Their combined value is over USD 180 billion, slightly greater than the gross domestic product (GDP) of Hungary and slightly smaller than that of New Zealand (Figure 1.8). While the importance of these companies for Russia's economy, domestic politics and foreign relations is great, the academic literature on them is small.

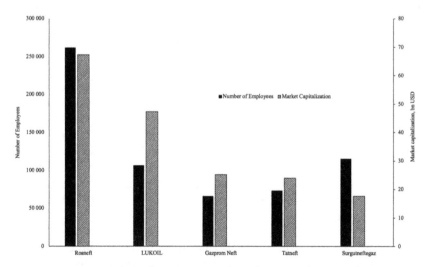

Figure 1.8   Number of employees and market capitalization of Russian oil companies (2018)

Source: Company websites.

Although oil-trading companies are also an important part of the Russian oil industry, we decided not to include them because some are subsidiaries of oil companies already covered in this book (for example, Litasco belongs to LUKOIL and Trumpet to Rosneft); they are mostly based outside Russia (Litasco in Switzerland, Trumpet in Ireland); employ a few people and information about them is difficult to obtain. Several companies have played important roles in the past but have been taken over by others. For example, Bashneft, Itera, Tyumen Oil Company (TNK) and YUKOS were all taken over by Rosneft, while Sibneft was taken over by Gazprom. Since this book aims to be current and forward-looking, we have chosen to focus on the current configuration of the companies rather than that of their past.

CORPORATE GOVERNANCE IN RUSSIAN COMPANIES

The governance structures of Russian companies often resemble those in the West in form but differ in content (Deloitte 2015; Kuznetsov and Kuznetsova 2009). To gain a deeper understanding of the companies discussed in this volume, it is worth highlighting some of the features of Russian corporate governance.

The general shareholder meeting constitutes the highest organ in the organizational structure of Russian companies (Tricker 2015, p. 306). Similar to companies in many Western countries, most Russian firms have a board of directors to govern the company. In practice, however, Russian boards of directors are often used as a mechanism for rubber-stamping the initiatives of influential company owners. Deals are presented to the boards for consideration at a late stage, and their job is seen as rejecting or (more likely) approving a deal but not getting involved in its details (Porshakov et al. 2010).

In a survey of company boards, Deloitte (2015) established that the average share of independent directors in Russian firms was only 43%, whereas the average for the UK was 61%, with 72%, 75% and 83% for Finland, the Netherlands and the United States, respectively. Moreover, in Russian companies, independent directors have limited access to corporate data and, therefore, little capacity to analyse the decisions they take.

In many Russian firms, in addition to the board of directors and the CEO, there is a management board in charge of the daily running of the firm. Members of the management board may make up as much as one-quarter of the board of directors, reducing the likelihood that the board will act as a check on the management (Teterevkova et al. 2017). These peculiarities of the structure of corporate governance in Russia facilitate practices such as share dilution, asset stripping, transfer pricing and ignoring the rights of minority shareholders (Adachi 2010; Porshakov et al. 2010).

Russian companies are also highly hierarchical and the stakes for employees are high. Mid- and top-level managers are often paid well, sometimes very well. However, should they run afoul of their superiors, they can be summarily dismissed. Disobedience and failure to carry out orders are often punished harshly, for example, through de facto bans on working for any company in the industry. Unsurprisingly, Russian employees are highly disciplined and obedient: great assets under good management but less so under bad management.

REFERENCES

Aalto, P. (2016), 'Modernisation of the Russian energy sector: Constraints on utilising Arctic offshore oil resources', *Europe-Asia Studies*, **68** (1), 38–63.

Adachi, Y. (2010), *Building Big Business in Russia: The Impact of Informal Corporate Governance Practices*, London: Routledge.

Appel, H. (2008), 'Is it Putin or is it oil? Explaining Russia's fiscal recovery', *Post-Soviet Affairs*, **24** (4), 301–23.

Arezki, R., V.A. Ramey and L. Sheng (2017), 'News shocks in open economies: Evidence from giant oil discoveries', *The Quarterly Journal of Economics*, **132** (1), 103–55.

Armitage, J. (2015), 'Igor Sechin: The oil man at the heart of Putin's Kremlin', accessed 3 December 2017 at http://www.independent.co.uk/news/business/analysis-and-features/igor-sechin-the-oil-man-at-the-heart-of-putins-kremlin-10043230.html.

Balmaceda, M. (2008), *Energy Dependency, Politics and Corruption in the Former Soviet Union: Russia's Power, Oligarchs' Profits and Ukraine's Missing Energy Policy*, New York: Routledge.

Balzer, H. (2005), 'The Putin thesis and Russian energy policy', *Post-Soviet Affairs*, **21** (3), 210–25.

Barysch, K. (2008), *Pipelines Politics and Power. The Future of the EU-Russia Energy Relations*, London: Centre for European Reform.

Bayulgen, O. (2014), *Foreign Investment and Political Regimes: The Oil Sector in Azerbaijan, Russia, and Norway*, Cambridge: Cambridge University Press.

Ben-Amar, W. and P. McIlkenny (2015), 'Board effectiveness and the voluntary disclosure of climate change information', *Business Strategy and the Environment*, **24** (8), 704–19.

Besio, C. and A. Pronzini (2014), 'Morality, ethics, and values outside and inside organizations: An example of the discourse on climate change', *Journal of Business Ethics*, **119** (3), 287–300.

Boersma, T. and C. Johnson (2012), 'The shale gas revolution: U.S. and EU policy and research agendas', *Review of Policy Research*, **29** (4), 570–6.

Bogdanova, E. (2015), 'Cross-cultural collaboration in contemporary Russia: Problems of contracting', *Journal of Social Policy Studies*, **13** (1), 123–36.

Boute, A. (2016), 'Off-grid renewable energy in remote Arctic areas: An analysis of the Russian Far East', *Renewable and Sustainable Energy Reviews*, **59** (1), 1029–37.

Bradshaw, M. (2006), 'Observations on the geographical dimensions of Russia's resource abundance', *Eurasian Geography and Economics*, **47** (6), 724–46.

Bulavka, L. and A. Buzgalin (2016), 'The oligarch, the state and the intelligentsia: Khodorkovsky as a mirror of the counterpoints of post-Soviet Russia', *Science & Society*, **80** (2), 248–56.

Campbell, D. (2013), 'After Doha: What has climate change policy accomplished?', *Journal of Environmental Law*, **25** (1), 125–36.

Casier, T. (2011), 'The rise of energy to the top of the EU-Russia agenda: From inter-dependence to dependence?', *Geopolitics*, **16** (3), 536–52.

Cheloukhine, S. and J. King (2007), 'Corruption networks as a sphere of investment activities in modern Russia', *Communist and Post-Communist Studies*, **40** (1), 107–22.

Clo, A. (2000), *Oil Economics and Policy*, New York: Springer.

Correljé, A. (2018), 'The Netherlands: Resource management and civil society in the natural gas sector', in Indra Overland (ed.), *Public Brainpower: Civil Society and Natural Resource Management*, Cham: Palgrave Macmillan, pp. 181–200.

Dahlgren, L. (2009), *IKEA Älskar Ryssland: En Berättelse Om Ledarskap, Passion Och Envishet*, Stockholm: Natur & Kultur.

Davydova, A. (2017a), 'Parnikovye gazy vpisyvayut v zakon', *Kommersant*, **35**, 2.

Davydova, A. (2017b), 'Russia wants to protect itself from climate change – without reducing carbon emissions', accessed 21 November 2017 at http://www.sciencemag.org/news/2017/09/russia-wants-protect-itself-climate-change-without-reducing-carbon-emissions.

de Graaff, N. (2012), 'Oil elite networks in a transforming global oil market', *International Journal of Comparative Sociology*, **53** (4), 275–97.

Deloitte (2015), 'Corporate governance structures of public Russian companies', accessed 25 November 2017 at https://www2.deloitte.com/ru/en/pages/risk/articles/2016/corporate-governance-structures-of-public-russian-companies.html.

Dixon, S. (2008), *Organisational Transformation in the Russian Oil Industry*, Cheltenham, UK and Northampton, MA, USA: Edward Elgar.

EIA (2019a), 'Natural gas gross withdrawals and production', accessed 13 October 2019 at https://www.eia.gov/dnav/ng/ng_prod_sum_dc_NUS_mmcf_a.htm.

EIA (2019b), 'Petroleum & other liquids', accessed 13 October 2019 at https://www.eia.gov/petroleum/data.php#crude.

EIA (2020), 'Europe Brent spot prices FOB', accessed 4 January 2020 at https://www.eia.gov/dnav/pet/hist/LeafHandler.ashx?n=PET&s=rbrte&f=M.

Elder, M. (2012), 'Gazprom feels the chill as its dominance is weakened', *Guardian*, accessed 3 December 2017 at http://www.theguardian.com/environment/2012/nov/15/gazprom-chill-shale-gas-revolution.

Fabry, N. and S. Zeghni (2002), 'Foreign direct investment in Russia: How the invest-ment climate matters', *Communist and Post-Communist Studies*, **35** (3), 289–303.

Farand, C. (2017), 'Vladimir Putin changes his mind and echoes Donald Trump to say humans are not to blame for climate change', accessed 29 November 2017 at http://www.independent.co.uk/news/world/europe/vladimir-putin-russia-climate-change-not-caused-by-humans-echoes-us-trump-a7660941.html.

Fjaertoft, D. and I. Overland (2015), 'Financial sanctions impact Russian oil, equip-ment export ban's effects limited', *Oil & Gas Journal*, **113** (8), 66–72, accessed 19 April 2020 at https://www.researchgate.net/publication/281776234_Financial_Sanctions_Impact_Russian_Oil_Equipment_Export_Ban's_Effects_Limited.

Fouquet, R. (2016), 'Historical energy transitions: Speed, prices and system transfor-mation', *Energy Research & Social Science*, **22** (C), 7–12.

Fouquet, R. and P.J. Pearson (2012), 'Past and prospective energy transitions: Insights from history', *Special Section: Past and Prospective Energy Transitions – Insights from History*, **50** (Suppl. C), 1–7.

Friedman, T. (2006), 'The first law of petropolitics', *Foreign Policy*, 28–36.

Gaddy, C. (2013), *Bear Traps on Russia's Road to Modernization*, New York: Routledge.

Gaddy, C. and B.W. Ickes (2019), *Russia's Addiction: How Oil, Gas, and the Soviet Legacy Have Shaped a Nation's Fate*, Washington, DC: Brookings Institution Press.

Gans-Morse, J. (2017), 'Demand for law and the security of property rights: The Case of post-Soviet Russia', *American Political Science Review*, **111** (2), 338–59.

Goldman, M. (2010), *Oilopoly: Putin, Power and the Rise of the New Russia*, Oxford: Oneworld Publications.

Grace, J.D. (2005), *Russian Oil Supply: Performance and Prospects*, Oxford: Oxford University Press.

Grealy, N. (2012), 'Gazprom as shale gas loser in denial', accessed 3 December 2017 at http://www.reimaginegas.com/?p=1991.

Griffin, P.A., A.M. Jaffe, D.H. Lont and R. Dominguez-Faus (2015), 'Science and the stock market: Investors' recognition of unburnable carbon', *Energy Economics*, **52**, 1–12.

Grubler, A. (2012), 'Energy transitions research: Insights and cautionary tales', *Special Section: Past and Prospective Energy Transitions – Insights from History*, **50** (C), 8–16.

Gustafson, T. (2012), *Wheel of Fortune: The Battle for Oil and Power in Russia*, Cambridge, MA: Belknap Press.

Henderson, J. and A. Ferguson (2014), *International Partnership in Russia – Conclusions from the Oil and Gas Industry*, Cham: Palgrave Macmillan.

Henderson, J. and A. Moe (2016), 'Gazprom's LNG offensive: A demonstration of monopoly strength or impetus for Russian gas sector reform?', *Post-Communist Economies*, **28** (3), 281–99.

Hendrix, C.S. (2015), 'Oil prices and interstate conflict', *Conflict Management and Peace Science*, **34** (6), 575–96.

Henry, L.A. and L. Sundstrom (2007), 'Russia and the Kyoto Protocol: Seeking an alignment of interests and image', *Global Environmental Politics*, **7** (4), 47–69.

Hiatt, S.R., J.B. Grandy and B.H. Lee (2015), 'Organizational responses to public and private politics: An analysis of climate change activists and U.S. oil and gas firms', *Organization Science*, **26** (6), 1769–86.

Hill, F. (2004), *Energy Empire: Oil, Gas and Russia's Revival*, London: Foreign Policy Centre.

Hufbauer, G.C. and J. Kim (2010), 'Reaching a global agreement on climate change: What are the obstacles?', *Asian Economic Policy Review*, **5** (1), 39–58.

IEA (2019), 'United States production', accessed 12 October 2019 at https://www.iea.org/ugforum/ugd/united%20states/.

Index Mundi (2020), 'Russian natural gas monthly price – US dollars per million metric British thermal unit', accessed 3 January 2020 at https://www.indexmundi.com/commodities/?commodity=russian-natural-gas&months=300.

Ivolga, A., A. Pominov, E. Sukhareva, S. Ilya and M. Shigreva (2018), *Transparency in Corporate Reporting: Assessing Russia's Largest Companies*, Moscow: Transparency International.

Jirušek, M., T. Vlček and J. Henderson (2017), 'Russia's energy relations in Southeastern Europe: An analysis of motives in Bulgaria and Greece', *Post-Soviet Affairs*, **33** (5), 335–55.

Kalehsar, O.S. and A. Telli (2017), 'The future of Iran–Russia energy relations post-sanctions', *Middle East Policy*, **24** (3), 163–70.

Kennaway, A. (2000), 'Collected writings', accessed 20 May 2018 at https://www.files
.ethz.ch/isn/43988/M20_Collected.pdf.

Kotkin, S. (2008), *Armageddon Averted: The Soviet Collapse, 1970–2000*, Oxford and
New York: Oxford University Press.

Kraemer, A. (2017), 'Can Russia imagine a post-fossil fuel future?', accessed
21 November 2017 at http://blog.iass-potsdam.de/2017/05/russia-post-fossil-fuel
-future/.

Kramer, A.E. (2011), 'Exxon wins prized access to Arctic with Russia deal', *New
York Times*, accessed 29 November 2017 at https://www.nytimes.com/2011/08/31/
business/global/exxon-and-rosneft-partner-in-russian-oil-deal.html.

Kuznetsov, A. and O. Kuznetsova (2009), 'Corporate governance in Russia: Concept
and reality', in R.W. McGee (ed.), *Accounting Reform in Transition and Developing
Economies*, Boston, MA: Springer US, pp. 445–57.

Kyj, M.J. and L.S. Kyj (2010), 'A relational framework for analyzing ventures in cogni-
tive environments: Illustrations from the TNK-BP experience', *Journal of East-West
Business*, **16** (4), 340–60.

Ledeneva, A. (2013), *Can Russia Modernise? Sistema, Power Networks and Informal
Governance*, Cambridge: Cambridge University Press.

Liuhto, K. (2010), 'Political risk for foreign firms in Russia', *Journal for Economic
Forecasting*, **3**, 141–57.

Locatelli, C. (2006), 'The Russian oil industry between public and private governance:
Obstacles to international oil companies' investment strategies', *Energy Policy*, **34**,
1075–85.

Lunden, L.P., D. Fjaertoft, I. Overland and A. Prachakova (2013), 'Gazprom vs. other
Russian gas producers: The evolution of the Russian gas sector', *Energy Policy*, **61**,
663–70.

Makarov, A.A., T.A. Mitrova and V.A. Malakhov (2013), 'Prognoz mirovoy energetiki
i posledstviya dlya Rossii', accessed 20 November 2017 at https://cyberleninka.ru/
article/v/prognoz-mirovoy-energetiki-i-posledstviya-dlya-rossii.

Makarov, A.A., T.A. Mitrova, L.M. Grigoryev et al. (2014), 'Prognoz razvitiya energetiki mira i Rossii do 2040 goda', Moscow: INEI RAN, accessed 3 December 2017 at
https://www.eriras.ru/files/forecast_2040.pdf.

Makarov, A.A., E.D. Belotskaya, F.V. Beselov et al. (2016), 'Prognoz razvitiya energetiki mira i Rossii 2016', accessed 20 November 2017 at http://ac.gov.ru/files/
publication/a/10585.pdf.

Makarov, A.A., V.V. Briliantova, Y.V. Galkin et al. (2019), 'Prognoz razvitiya energetiki mira i Rossii 2019', accessed 20 November 2019 at https://energy.skolkovo.ru/
downloads/documents/SEneC/Research/SKOLKOVO_EneC_Forecast_2019_Rus
.pdf.

Maury, B. and E. Liljeblom (2009), 'Oligarchs, political regime changes, and firm
valuation', *Economics of Transition*, **17** (3), 411–38.

Meadowcroft, J. (2009), 'What about the politics? Sustainable development, transition
management, and long-term energy transitions', *Policy Sciences*, **42** (4), 323–40.

Meredith, S. and G. Cutmore (2017), 'Climate change doubters may not be so silly, says
Russia's President Putin', accessed 29 November 2017 at https://www.cnbc.com/
2017/03/30/vladimir-putin-russia-trump-us-climate-policy.html.

Mironov, V. and A. Petronevich (2015), 'Discovering the signs of Dutch disease in
Russia', *Resources Policy*, **46**, 97–112.

Mitchell, J.V. and B. Mitchell (2014), 'Structural crisis in the oil and gas industry',
Energy Policy, **64** (C), 36–42.

Novak, A. (2016), 'Intervyu Aleksandra Novaka radiostantsii "Ekho Mosvky"', accessed 21 November 2017 at https://minenergo.gov.ru/node/6142.

O'Sullivan, M., I. Overland, D. Sandalow et al. (2017), 'The geopolitics of renewable energy', accessed 22 November 2017 at https://www.researchgate.net/publication/317954274.

Obydenkova, A. and A. Libman (2015), 'Understanding the survival of post-Communist corruption in contemporary Russia: The influence of historical legacies', *Post-Soviet Affairs*, **31** (4), 304–38.

Overland, I. (2011), 'Modernization after Medvedev?', *Russian Analytical Digest*, **105**, 2–4.

Overland, I. (2015), 'Future petroleum geopolitics: Consequences of climate policy and unconventional oil and gas', in Jinyue Yan (ed.), *Handbook of Clean Energy Systems*, Chichester: John Wiley, pp. 3517–44, https://doi.org/10.1002/9781118991978.hces203.

Overland, I. (2018), 'Norway: Public debate and the management of petroleum resources and revenues', in Indra Overland (ed.), *Public Brainpower: Civil Society and Natural Resource Management*, Cham: Springer International Publishing, pp. 217–45.

Overland, I. and H. Kjaernet (2009), *Russian Renewable Energy: The Potential for International Cooperation*, Aldershot: Ashgate.

Overland, I. and G. Kubayeva (2018), 'Did China bankroll Russia's annexation of Crimea? The role of Sino-Russian energy relations', in H. Blakkisrud and E. Wilson Rowe (eds), *Russia's Turn to the East: Domestic Policymaking and Regional Cooperation*, Cham: Palgrave Macmillan, pp. 95–118.

Overland, I., J. Godzimirski, L.P. Lunden and D. Fjaertoft (2013), 'Rosneft's offshore partnerships: The re-opening of the Russian petroleum frontier?', *Polar Record*, **49** (2), 140–53.

Overland, I., A. Bambulyak, A. Bourmistrov, O.T. Gudmestad, F. Mellemvik and A. Zolotukhin (2015), 'Barents Sea oil and gas 2025: Three scenarios', in O.T. Gudmestad, I. Overland and A. Zolotukhin (eds), *International Arctic Petroleum Cooperation: Barents Sea Scenarios*, London: Routledge, pp. 11–32.

Peng, M. (2000), *Business Strategies in Transition Economies*, London: Sage.

Person, R. (2016), 'The deep impact of economic collapse on democratic support', *Problems of Post-Communism*, **63** (5–6), 335–53.

Porshakov, S., C. Gilbert, A. Ivakhnik and E. Chumakova (2010), 'Modern corporate governance in Russia as seen by foreign businessmen and experts', accessed 25 November 2017 at https://www2.deloitte.com/content/dam/Deloitte/ru/Documents/finance/modern_corporate_governance_russia.pdf.

Poussenkova, N. and I. Overland (2018), 'Russia: Public debate and the petroleum sector', in Indra Overland (ed.), *Public Brainpower: Civil Society and Natural Resource Management*, Cham: Springer International Publishing, pp. 261–89.

Proedrou, F. (2017), 'Revisiting pipeline politics and diplomacy', *Problems of Post-Communism*, **65** (6), 409–18.

Putin, V. (2017a), 'Gastbeitrag von Wladimir Putin: "Wir teilen Die Deutschen prioritäten"', accessed 21 November 2017 at http://www.handelsblatt.com/my/politik/international/gastbeitrag-von-wladimir-putin-wir-teilen-die-deutschen-prioritaeten/20020620.html.

Putin, V. (2017b), 'O chem rasskazal Vladimir Putin ya plenarnom zasedanii PMEF', *Rossiyskaya Gazeta*, accessed 20 November 2017 at https://rg.ru/2017/06/02/reg-szfo/o-chem-rasskazal-vladimir-putin-na-plenarnom-zasedanii-pmef.html.

Reynolds, D.B. and M. Kolodziej (2008), 'Former Soviet Union oil production and GDP decline: Granger causality and the multi-cycle Hubbert curve', *Energy Economics*, **30** (2), 271–89.

Rivera, D.W. and S.W. Rivera (2014), 'Is Russia a militocracy? Conceptual issues and extant findings regarding elite militarization', *Post-Soviet Affairs*, **30** (1), 27–50.

Rutland, P. (2015), 'Petronation? Oil, gas, and national identity in Russia', *Post-Soviet Affairs*, **31** (1), 66–89.

Schlichting, I. (2013), 'Strategic framing of climate change by industry actors: A meta-analysis', *Environmental Communication*, **7** (4), 493–511.

Sevastyanov, S. (2008), 'The more assertive and pragmatic new energy policy in Putin's Russia: Security implications for Northeast Asia', *East Asia*, **25** (1), 35–55.

Sharmina, M. (2017), 'Low-carbon scenarios for Russia's energy system: A participative backcasting approach', *Energy Policy*, **104** (C), 303–15.

Sharples, J.D. (2016), 'The shifting geopolitics of Russia's natural gas exports and their impact on EU-Russia gas relations', *Geopolitics*, **21** (4), 880–912.

Shaukat Khan, T., T.T.T. Nguyen, F.L. Ohnsorge and R. Schodde (2016), 'From commodity discovery to production', *The World Bank*, pp. 1–23.

Skryzhevska, Y., V.-P. Tynkkynen and S. Leppänen (2015), 'Russia's climate policies and local reality', *Polar Geography*, **38** (2), 146–70.

Smil, Vaclav (2010), *Energy Transitions: History, Requirements, Prospects*, Santa Barbara: ABC-CLIO.

Smith, N. and E. Thomas (2015), 'Determinants of Russia's informal economy: The impact of corruption and multinational firms', *Journal of East-West Business*, **21** (2), 102–28.

Sovacool, B.K. (2017), 'Experts, theories, and electric mobility transitions: Toward an integrated conceptual framework for the adoption of electric vehicles', *Energy Research & Social Science*, **27**, 78–95.

Spencer, B. and B. Hansen (2012), 'Russia's Gazprom skeptical of US-led shale gas boom', accessed 3 December 2017 at https://www.platts.com/latest-news/natural -gas/washington/feature-russias-gazprom-skeptical-of-us-led-shale-8545823.

Sprinz, D.F., B.B. de Mesquita, S. Kallbekken, F. Stokman, H. Sælen and R. Thomson (2016), 'Predicting Paris: Multi-method approaches to forecast the outcomes of global climate negotiations', *Politics and Governance*, **4** (3), 172–87.

Stulberg, A.N. (2012), 'Strategic bargaining and pipeline politics: Confronting the credible commitment problem in Eurasian energy transit', *Review of International Political Economy*, **19** (5), 808–36.

Supran G and N. Oreskes (2017), 'Assessing ExxonMobil's climate change communications (1977–2014)', *Environmental Research Letters*, **12** (8), 1–18.

TASS (2017), 'Minenergo: Zapadnye sanktsii ne povliyali na potok investitsii na shelf RF', accessed 20 November 2017 at http://tass.ru/ekonomika/4504363.

TASS (2018), 'Eksperty: Zavisimost ekonomiki RF i budzheta ot nefti snova nachala vozrastat', accessed 30 May 2018 at http://tass.ru/ekonomika/4941082.

Teterevkova, E., S. Fedorov and V. Gareev (2017), 'Corporate governance and directors' duties: Russian Federation', accessed 25 November 2017 at https:// uk.practicallaw.thomsonreuters.com/5-502-1245?transitionType=Default& contextData=(sc.Default)&firstPage=true&bhcp=1.

Tompson, W. (2005), 'The political implications of Russia's resource-based economy', *Post-Soviet Affairs*, **21** (4), 335–59.

Tricker, R.I. (2015), *Corporate Governance: Principles, Policies, and Practices*, Oxford: Oxford University Press.

Tynkkynen, V.-P. and N. Tynkkynen (2018), 'Climate denial revisited: (Re)contextu-alising Russian public discourse on climate change during Putin 2.0', *Europe-Asia Studies*, **70** (7), 1103 20.

Valeeva, V. (2017), 'Stop the blame game: Russia is waking up to climate change', accessed 21 November 2017 at http://blog.iass-potsdam.de/2017/09/stop-blame -game-russia-waking-climate-change/.

van der Ploeg, F. (2016), 'Fossil fuel producers under threat', *Oxford Review of Economic Policy*, **32** (2), 206–22.

Walker, C. (2015), 'Stability and precarity in the lives and narratives of working-class men in Putin's Russia', *Social Alternatives*, **28** (34), 28–34.

Wigell, M. and A. Vihma (2016), 'Geopolitics versus geoeconomics: The case of Russia's geostrategy and its effects on the EU', *International Affairs*, **92** (3), 605–27.

Wright, P. and J.C. Boué (2018), 'The United Kingdom: Public debate and the man-agement of petroleum resources', in Indra Overland (ed.), *Public Brainpower: Civil Society and Natural Resource Management*, Cham: Palgrave Macmillan, pp. 329–46.

Wright, S. (2017), 'Qatar's LNG: Impact of the changing East-Asian market', *Middle East Policy*, **24** (1), 154–65.

2. Rosneft: lord of the rigs[1]

Rosneft is Russia's national oil company. Its evolution mirrors that of the relationship between the state and oil corporations in post-Soviet Russia. Rosneft was born from the ashes of the Soviet state structures in the early 1990s but went into a seemingly terminal decline as the feisty oligarchs hacked away at its assets during this chaotic decade. However, the rise of Vladimir Putin and his entourage gave Rosneft a new lease of life, and in the 2000s, it came to be the Tyrannosaurus rex of the Russian petroleum sector. It grew mainly through acquisitions, beginning with Yuganskneftegaz and YUKOS in 2004 and 2007, respectively. With Igor Sechin as chairman of its board from 2004 and then President from 2012, the company bought three of Russia's other major oil and gas companies – TNK-BP, Itera and Bashneft – becoming the world's largest publicly traded oil company by hydrocarbon reserves and production. This history of rise and fall and rise again contrasts with that of Russia's other major state company, Gazprom. Gazprom's position as the blue whale of the Russian gas sector during this period was far more consistent and stable.

In 2018, Rosneft produced 230 million tonnes of oil and condensate, up 2.1% from 2017. That year, its gas production amounted to 67 billion cubic metres. Its proved reserves of hydrocarbons grew by 4% to 41 billion barrels of oil equivalent. Rosneft is also the largest Russian refiner, handling 103 million tonnes of oil in 2018 (Rosneft 2019b). It controls 13 major refineries, three petrochemical companies and four gas-processing plants. It also owns shares in three German refineries and a stake in the Mozyr refinery in Belarus. Its petroleum product distributors operate in 66 regions of Russia and other post-Soviet states. As of December 2016, the company had a network of 2962 gas stations.

Rosneft contributed to transforming Russian foreign energy policy by turning to Chinese and Indian partners, driven by both commercial motives and political necessity. Rosneft actively established strategic partnerships with international oil companies both in Russia and abroad. Before the imposition of the Western sanctions in connection with the conflict in Ukraine, Rosneft was engaged in developing shale oil, Arctic offshore fields and other parts of Russia's continental shelf.

Rosneft invests significant effort in monitoring and forecasting oil prices and macroeconomic trends. However, given its strong political ties, it is also well aware that the state will pitch in to help it cope with unforeseen developments.

As of April 2018, Rosneft's main shareholder was the state-owned vehicle Rosneftegaz, which held 50.00000001% of the company's stock (Table 2.1). In November 2019, the company's market capitalization was USD 74 billion, less than those of Chevron, Equinor or ExxonMobil but greater than ENI, Gazprom or LUKOIL (Bloomberg 2019).

Table 2.1   *Rosneft's shareholders*

Shareholder	% of stock
Rosneftegaz	50.00000001
BP	19.75
QHG Oil Ventures	19.50
National Settlement Depository	10.38
Federal State Property Agency	One share

Source: Rosneft website.

CORPORATE HISTORY: EVOLVING ROSNEFT IN AN EVOLVING RUSSIA

The 1990s: The Birth and Near-Death Experience of Rosneft

Rosneft originated in the state corporation Rosneftegaz, which was established in October 1991 on the basis of the dismantled USSR Ministry for the Oil Industry. Presidential Decree 1403 of 17 November 1992[2] created vertically integrated oil companies LUKOIL, YUKOS and Surgutneftegas, the state enterprise Rosneft, as well as the Transneft and Transnefteproduct transportation companies. Rosneft was expected to manage government stakes in 259 (out of 301) oil industry enterprises and perform additional non-commercial functions. These included supporting the restructuring of the joint stock companies (JSCs), ensuring deliveries of fuel to cover state needs, coordinating government investments in the sector and organizing the manufacture of oilfield equipment (Hudson and Poussenkova 1996). At the time, Rosneft was headed by Alexander Putilov, the former General Director of Uraineftegaz – now a LUKOIL subsidiary.

In 1993, Rosneft accounted for more than 60% of Russian oil production. Given all the other problems of the transition period, the Russian state struggled to manage this giant conglomerate. The company's complicated organizational structure, including 26 oil-producing associations, 23 refineries, several gas-processing plants and several petroleum product suppliers and research institutes, hardly facilitated the management of the company.

Between 1993 and 1995, several new vertically integrated oil companies were carved out of the assets controlled by Rosneft: Slavneft, SIDANCO, Sibneft,

TNK, Eastern Oil Company, ONACO and KomiTEK. As a result, Rosneft kept shrinking. The creation of Sibneft, which included Noyabrskneftegaz (a 22-million-tonne-per-annum oil producer) and Omsknefteorgsyntez (the most advanced refinery in Russia), hit Rosneft particularly hard.

Presidential Decree 327 of 1995 transformed the state enterprise Rosneft into an open joint stock company (OJSC).[3] However, the losses continued: Rosneft was deprived of a stake in the Moscow refinery and Mosnefteprodukt, which it had held in trust management (Neft i kapital 2004a, p. 14). In 1997, the government decided to sell Krasnodarnefteorgsyntez because of tax debts of USD 32 million, and it was eventually auctioned off in 2000 (NGFR 2008).

Rosneft also nearly lost Purneftegaz – its key oil-producing subsidiary based in Yamal-Nenets Autonomous District (YNAO). When SIDANCO was established in May 1994, Purneftegaz was transferred to it. In early 1995, the leadership of Purneftegaz asked the government to return it to Rosneft, and it was handed back, but SIDANCO initiated a lawsuit to challenge the decision (Latyshova 1995, pp. 46–51). This tug of war continued for two years, but when the courts eventually recognized SIDANCO's claims, Vladimir Potanin, the owner of SIDANCO, unexpectedly returned it to Rosneft (Neft i kapital 2004b, p. 60).

Having lost valuable assets throughout the 1990s, Rosneft ended up as a minor player, accounting for only 7% of Russia's crude output and reserves. However, it managed to remain a tempting target for the oligarchs. Several attempts were made to privatize Rosneft during the 1990s, but the company remained under state patronage. In April 1997, Alexander Putilov was dismissed as Rosneft's President because he resisted the State Property Committee's plans to privatize the company (NGFR 2008). Since he still had influential friends, he was made the chairman of the board of directors, whereas Yuri Bespalov, representing Boris Berezovskiy – the most powerful oligarch of the 1990s – became President. Bespalov started preparing the company to be bought by Sibneft.

Attempts to privatize Rosneft in 1997 failed because influential players had an interest in delaying the process. The most controversial attempt was made in 1998. Just before his resignation, Prime Minister Victor Chernomyrdin approved a plan to sell 75% + 1 share of Rosneft for USD 2.1 billion plus USD 400 million to be invested in future projects to be implemented by the new owner of Rosneft. The auction was scheduled for 29 May 1998, and a fierce battle was expected between the competing alliances of SIDANCO-BP, Gazprom-Shell-LUKOIL and Yuksi (YUKOS-Sibneft) (Samoilova 1998). However, the May auction was a fiasco since the foreign investors stayed away, and the sale was postponed until 30 October. The sale price was reduced to USD 1.6 billion and the investment conditions to USD 65 million (Poluektov 1998). The government fired both Putilov and Bespalov and hired the Alliance Group led by Ziya Bazhayev, the former President of SIDANCO, to prepare Rosneft for the auction. By this time, Rosneft was in a desperate situation with large debts, unmanageable subsidiaries and 40 enterprises whose assets had been seized. Yet in just two weeks, Bazhayev and his team significantly enhanced Rosneft's investment appeal through restructuring. However, on 17 August 1998, the worst financial crisis in Russian history erupted, and the auction was postponed indefinitely.

In October 1998, Sergei Bogdanchikov, the former Head of Sakhalinmorneftegaz, was appointed President of Rosneft. He faced a mission impossible: to restore a company that had almost entirely disintegrated. Bogdanchikov immediately started setting his house in order. He appointed loyal people to run Rosneft's key production units Purneftegaz and Sakhalinmorneftegaz and made it mandatory for the managers of subsidiaries to obtain his personal approval of any hiring and firing, business plans and loans and government relations. He also forced them to channel all oil exports through the Rosneft holding (Neft i kapital 2004c, p. 161).

The 2000s: The Rebirth of Rosneft

When Sergei Bogdanchikov became the President of Rosneft, nobody expected that he would hold on to this position for long, as he was neither a political heavyweight nor a protégée of the oligarchs. However, he soon found a powerful sponsor at the very top – President Vladimir Putin – who needed a mighty state oil company to counterbalance the voracious private corporations. Thus, the year 2000 marked the start of a new chapter in Rosneft's life.

Sergei Bogdanchikov started by consolidating Rosneft's control over its subsidiaries. In 2000, the government allowed Rosneft to increase its stake in its subsidiaries to 75%, and it began to buy their shares on the secondary market. The renaissance of Rosneft was primarily accomplished through aggressive acquisitions, beginning in 2003 with Severnaya Neft with 17 oilfields in the Timan-Pechora province (Derbilova and Tutushkin 2004).

The rising status of Rosneft was confirmed in July 2004 when Igor Sechin, Deputy Head of the Presidential Administration and one of the most influential people in Russia, was made the chairman of Rosneft's board of directors, ensuring strong political patronage for the company. Rosneft became a national champion: owing to its large hydrocarbon reserves and political connections, it was expected to compete against the global majors, protect the national petroleum wealth and implement the government's energy policy.

Rosneft joined the big league in 2004 when it acquired Yuganskneftegaz – the key producing subsidiary of YUKOS – after Gazprom failed to secure this prize asset. In the early 2000s, before it was acquired by Rosneft, Yuganskneftegaz demonstrated double-digit crude output growth, mainly owing to the intensification of oil production practised by YUKOS and the commissioning of the Priobskoye field. When the government began the crackdown on YUKOS in the autumn of 2003, YUKOS was hit with a tax bill of USD 28 billion for the period 1999–2003 and Yuganskneftegaz was subsequently put up for sale (Poussenkova and Overland 2018).

It was widely expected that Gazprom would snap up Yuganskneftegaz. However, YUKOS sought court protection in Houston under Chapter 11 of the US law on bankruptcy. The court vetoed the Yuganskneftegaz auction and forbade Gazprom and six international banks from participating in the auction. The bankers obeyed, and Gazprom lost the chance of raising the foreign loans it needed to carry out the transaction. Despite this decision, the auction took place on 19 December 2004: the unknown company Baikal Finance Group, registered on the eve of the auction in the provincial town of Tver, bought Yuganskneftegaz for USD 9.35 billion (Mescherin 2005, p. 5). Three days later, Rosneft purchased Baikal Finance Group for a trifling RUB 10 000.

Having bought Yuganskneftegaz, Rosneft was transformed from a second-tier player with 21 million tonnes of oil production per annum into a 75-million-tonne-per-annum giant. At the same time, Rosneft became Russia's second-largest borrower after Gazprom. In early 2005, its debt amounted to USD 22.5 billion, and the rating agency Standard and Poor's (S&P) downgraded its credit rating from B to B- (Vinogradova 2005, p. 29).

It is worth noting that at the same time, in 2004, Rosneft was once again in danger of losing its independence. In September 2004, President Putin had approved a government proposal to incorporate Rosneft into the structure of Gazprom in exchange for 10.74% of its shares that were held by its subsidiaries. This would have permitted the state to increase its stake in Gazprom and to liberalize the trade in its shares. However, Gazprom realized that the reputational risks were too high. Moreover, Rosneft's value drastically increased after the purchase of Yuganskneftegaz, placing it beyond Gazprom's reach. Instead, Gazprom bought 72.6% of Sibneft's shares for USD 13 billion in September 2005 (see the chapter on Gazprom Neft in this book).

Ultimately, the government changed its plans diametrically. It transferred 100% of Rosneft's shares – then valued at USD 26 billion – to the ownership vehicle Rosneftegaz, which subsequently raised the necessary debt to purchase 10.74% of Gazprom's shares. After that, Rosneftegaz organized an initial public offering of Rosneft shares to repay the debt.

To raise money for the purchase of Yuganskneftegaz, Rosneft generated USD 6.1 billion by selling short-term bonds, and received another USD 1.8 billion of credit from Sberbank (Derbilova and Kudinov 2005). Vnesheconombank bought short-term bonds worth USD 5.3 billion from Rosneft using funds from the Ministry of Finance, which had been intended to repay Russia's external debt. In early 2005, Vnesheconombank raised USD

6 billion for Rosneft from Chinese banks. In exchange, Rosneft pledged to deliver 48 million tonnes of oil to CNPC up to 2010 (Derbilova 2006). With support from the Head of State and a powerful political figure on its board of directors, Rosneft had access to ministerial finances and carried out energy diplomacy on behalf of the Russian state.

It is paradoxical that during the 1990s, when broad privatization was ongoing in Russia, Rosneft remained under the wing of the state, but in the 2000s, when the étatization of the oil sector began, the company was partially privatized. Its initial public offering (IPO) took place in 2006: 14.8% of the shares were sold for USD 10.4 billion. The share price was USD 7.55, corresponding to a capitalization of USD 79.8 billion. Rosneft's IPO was the largest ever in Russia and the fifth largest in world history. Russian oligarchs Roman Abramovich, Oleg Deripaska and Vladimir Lisin paid USD 1 billion for stakes in Rosneft, thus making important political investments. BP, CNPC (China) and Petronas (Malaysia) bought shares worth USD 1 billion, USD 1.1 billion and USD 500 million, respectively. In addition, a people's IPO was organized for the citizens of Russia. As a result, 115 000 Russians became shareholders of Rosneft, spending USD 750 000 in total (Derbilova and Surzhenko 2006).

The bankruptcy of YUKOS in 2007 was another important milestone in Rosneft's evolution. Rosneft competed with Gazprom for the remaining assets of YUKOS, and the two state companies managed to obtain what they wanted: mainly oil assets in the case of Rosneft and gas assets in the case of Gazprom.

A scandal erupted during the sale of the YUKOS oil assets. In May 2007, the most expensive lot was sold, consisting of the eastern assets of YUKOS: Tomskneft, Vostsibneftegaz, the Angarsk Petrochemical Company and the Achinsk refinery. A Rosneft subsidiary, Neft-Aktiv, bought the lot at a price just above the starting price of USD 6.8 billion (Surzhenko and Mazneva 2007). However, soon Rosneft announced that it would sell a 50% stake in Tomskneft to Vnesheconombank for USD 3.4 billion. Analysts believed that Vnesheconombank planned to purchase this stake for Gazprom Neft. In the summer of 2007, Rosneft announced that the deal was closed, but Vladimir Dmitriev, Head of Vnesheconombank, denied that such a deal had taken place, while the bank's supervisory board stated that it had not reviewed the transaction (Petrachkova and Derbilova 2007). The scandal was hushed up, and in December 2007 Gazprom Neft acquired half of Tomskneft directly for USD 3.6 billion (Petrachkova and Derbilova 2007).

Rosneft won five out of 12 YUKOS bankruptcy auctions. Through these acquisitions, it improved its production/refining ratio and became the number one player in eastern Russia. It also became the world's largest public oil company by liquid hydrocarbon reserves (Surzhenko and Mazneva 2007). However, its debt rose to USD 36 billion, and, as a precautionary measure, in 2007, the government added Rosneft to its list of strategic enterprises that could be bankrupted only under a special procedure (Mazneva and Derbilova 2007).

Rosneft further strengthened its position in 2008 when Igor Sechin, the chairman of its board of directors, became Deputy Prime Minister with special responsibility for the fuel and energy sector. In May 2008, S&P upgraded its long-term credit rating from BB+ to BBB- (stable forecast), reflecting the benefits of Sechin's new position for Rosneft.

Combined with the financial crisis of 2008, the debts that Rosneft had incurred due to the acquisition of YUKOS forced it to seek out Chinese capital again. In February 2009, after hard bargaining with the Chinese and Igor Sechin personally leading the negotiations, Rosneft received a loan worth USD 15 billion from China. At the same time, the Chinese banks granted USD 10 billion to Transneft, mainly for the construction of the ESPO pipeline, including a spur to China. In return, the two state companies pledged to deliver 15 million tonnes per annum to China for 20 years (see also the section on internationalization below), with Vankor field playing a key role in these supplies.

The commissioning of the Vankor oil, gas and condensate field in 2009 is one of Rosneft's greatest triumphs. Located in Krasnoyarsk Krai, Vankor is the largest field discovered and launched in Russia in a quarter of a century. Vankor is actually a cluster of fields, with proved oil reserves exceeding 500 million tonnes and gas reserves of 182 billion cubic metres. Rosneft acquired Vankorneft in 2004, wresting it from YUKOS and Total.

Vankor was expected to reach a production of 25 million tonnes per annum at peak. However, the field peaked at 22 million tonnes per annum in 2014, staying at this level for three years and accounting for 11–12% of Rosneft's output. Production then began to fall, and there were fears that this trend might accelerate and reduce the output to 13 million tonnes by 2020 (Fadeeva 2016d). Rosneft anticipated that between 2016 and 2020, the development of the Suzunskoye, Tagulskoye and Lodochnoye fields of the Vankor cluster would offset the decline.

Post-2010: Further Expansion

In June 2010, Sergei Bogdanchikov's directorship expired, and there was another change of guard at Rosneft. Rumours had long been circulating about Bogdanchikov's possible dismissal mainly because of his troubled relationship with Sechin (Derbilova and Reznik 2010). In September 2010, President Dmitriy Medvedev, who had long been connected with Rosneft's rival Gazprom, appointed Eduard Khudainatov, Head of Gazprom subsidiary Severneftegazprom, to the presidency of Rosneft. Rosneft's acquisition spree continued under Khudainatov; it bought Taas-Yuryakh Neftegazodobycha, for example, and established strategic alliances with foreign majors – both trends likely masterminded by Igor Sechin.

In May 2012, just after the re-election of Putin as Russian President, the next phase of Rosneft's development was launched: the 'Rosneftization' of the Russian oil sector. Igor Sechin became the company's CEO and embarked on a new wave of major acquisitions.

In March 2013, Rosneft acquired TNK-BP. It bought BP's 50% stake in the company for USD 16.65 billion and 12.84% of Rosneft's shares, while BP purchased 5.66% of Rosneft's shares from Rosneftegaz, becoming the second-largest shareholder of Rosneft after the state with 19.75% (it had bought 1.25% during Rosneft's IPO). Rosneft also bought a stake of the Alfa-Access-Renova (AAR) consortium in TNK-BP for USD 27.73 billion. Rosneft raised USD 51 billion for the purchase, including USD 31 billion of credits from foreign banks (Solodovnikova 2013). As a result of the acquisition, Rosneft was transformed into the world's largest public oil company in terms of both hydrocarbons production and reserves (Rosneft 2013a). Rosneft benefited by obtaining several of TNK-BP's international projects and attractive gas assets as well as prolific reserves in Western Siberia (such as the legendary Samotlor field[4]) and Eastern Siberia.

Rosneft's acquisition of TNK-BP inspired a new round of China-Russia oil cooperation in 2013. Two major oil export contracts were signed in 2013. The first contract was between Rosneft and CNPC for the delivery of 360 million tonnes of crude over 25 years and was worth USD 270 billion with a stipulated advance payment of USD 65 billion. The second contract was with Sinopec and worth USD 85 billion for 100 million tonnes of crude over ten years. This enabled the acquisitions spree to continue with the purchase of the independent gas producer Itera, oil assets of ALROSA and Sibneftegaz (Rosneft 2013b, 2013c; Stulov 2016a).

In October 2016, the federal government found that it urgently needed money for the state budget and wanted to raise it through the privatization of several companies. Initially, there were plans to hold a public auction for Bashneft – a company formerly owned by the regional government of Bashkortostan – which had been privatized in 2002–03 but then seized by the Russian government in 2014. LUKOIL, NNK (Nezavisimaya Neftegazovaya Kompaniya of Eduard Khudainatov) and other players were interested in Bashneft's assets. Some government officials who were against Rosneft's participation in this auction argued that it would not add money to the state coffers if one state company bought another (Fadeeva et al. 2016a). But Igor Sechin actively lobbied for Rosneft's interests, and ultimately it turned out to be the only contender for the stake. Through this acquisition, Rosneft's production of liquids grew by 10% and oil refining by 20%. Sechin argued that the acquisition of Bashneft would raise the capitalization of Rosneft, making it possible to subsequently privatize 19.5% of Rosneft for over USD 11 billion (Stulov 2016b).

While Rosneft's close ties to the Kremlin often helped it achieve its objectives, in some cases they proved detrimental to the company. In the summer of 2014, following the conflict in Ukraine, the United States imposed personal sanctions against Igor Sechin and financial sanctions against Rosneft; Canada and the European Union (EU) also introduced sanctions against Rosneft (Fjaertoft and Overland 2015; Politinformatsia 2015). In addition, Rosneft was affected by the sectoral sanctions specifically targeting the Arctic, deep-water and shale projects. Moreover, in 2015, Rosneft's subsidiaries were also added to the US sanctions list (Usov et al. 2015). Igor Sechin complained that the inclusion of Rosneft on the sanctions list was 'unsubstantiated, subjective and illegal because the company played no role in the Ukrainian crisis' (Newsru.com 2014).

The next phase of Rosneft's privatization took place in 2016. When the government drew up plans to continue the privatization of the company to replenish the state coffers in the mid-2010s, Igor Sechin was against the idea. He argued that it would be more efficient to sell 19.5% of Rosneft when oil prices rebounded to the level of USD 100 per barrel (Fadeeva 2016a). However, other government officials insisted on the further partial privatization of Rosneft, overcoming his objections. Experts predicted that either CNPC would buy a 19.5% stake in Rosneft or Rosneft would find a way to buy itself (Barsukov et al. 2016). In autumn 2016, Igor Sechin held negotiations with some 30 potential investors. Finally, to everyone's surprise, a 50:50 consortium between Glencore and the Qatar Investment Authority purchased the stake for EUR 10.2 billion (less than the government had expected) (Fadeeva et al. 2016b). Rosneftegaz covered the difference by issuing additional dividends of RUB 18 billion. The consortium financed the deal from its own funds (EUR 2.8 billion) and bank funds (EUR 7.4 billion), mainly from Intesa Sanpaolo. Later, information leaked out that Gazprombank and VBRR (All-Russian Bank for Regional Development) had also helped the consortium to raise money (Stulov 2016c).

In 2017, Igor Sechin suddenly found another potential investor for Rosneft, the largely unknown Chinese company CEFC, which was expected to buy the lion's share of the stake held by the consortium. However, these plans did not materialize after CEFC's Head was detained in China on suspicion of economic crimes in early 2018 (Petleva 2018).

Due to its wide range of privatization schemes and acquisitions, Rosneft has a highly internationalized board of directors. From September 2017 onwards, seven out of 11 directors were foreigners; the chairman was also a foreigner – the former German Chancellor Gerhard Schroeder. There were also three foreigners on the 11-member board.

COMPANY PROFILE

As a result of its many acquisitions, Rosneft has a highly diverse portfolio of oil and gas assets, consisting of very mature oilfields in the centre of the European part of Russia, mature fields in Western Siberia, relatively young fields in Western Siberia and Timan-Pechora, young fields in Eastern Siberia, offshore producing fields in the Sakhalin-1 project and untapped reserves in the Arctic seas.

Offshore

Rosneft has always made an effort to develop its offshore operations. However, the results have been mixed, mainly because of geological issues and domestic and international economics and politics. Rosneft started to expand its Arctic offshore operations in the early 2000s. In 2001, Gazprom and Rosneft established the joint venture Sevmorneftegaz to develop several northern fields, including Kharampurskoye, Prirazlomnoye and Shtokmanovskoye. By 2002, Rosneft had gained de facto control over the projects of Sevmorneftegaz, and it was thought that its central role in the development of the Prirazlomnoye field could enable the company to take on a leading role on the Russian Arctic continental shelf (Neft i kapital 2003, p. 24). However, in 2005, Rosneft had to sell its stake in the joint venture to Gazprom to pay for the acquisition of YUKOS's former crown jewel – Yuganskneftegaz.

In the early 2000s, Rosneft also consolidated its position on Sakhalin by launching other offshore projects in addition to Sakhalin-1, in which it had been involved since 1996, when the relevant production-sharing agreement (PSA) was signed. In 2002, together with BP, it received the geological licences for Sakhalin-4 and Sakhalin-5. BP was to finance the exploration. However, both projects failed: the first two wells drilled on the Zapadno-Schmidtovskiy block were dry, and in 2009, the partners gave up Sakhalin-4. In 2007, they discovered a field within Kaigansko-Vasyukanskiy block, but BP considered it commercially unattractive and withdrew from Sakhalin-5 in 2011 (Gavshina 2011).

In August 2003, Rosneft won a five-year exploration licence for the Western Kamchatka continental shelf and signed a memorandum of understanding with the Korean National Oil Corporation (KNOC) in September 2004. Again, the foreigners were to cover the exploration expenses. However, in August 2008, Rosnedra (Federal Agency for Subsoil Use) refused to extend the exploration licence because of non-fulfilment of the drilling plan by Rosneft, and this asset was transferred to Gazprom as the company responsible for the gasification of Kamchatka.

Rosneft has also been active in the Sea of Azov and the Black Sea since 2002. Together with LUKOIL and the Administration of Krasnodar Krai, Rosneft established Priazovneft to assess and develop the reserves of the Sea of Azov. In spring 2003, Rosneft and Total signed an agreement on the joint exploration of the Black Sea, including the Tuapse Trough. However, their partnership did not last, mainly because of the controversies surrounding Vankor.

Under an important piece of legislation passed in 2008, only Gazprom and Rosneft were eligible to work on Russia's continental shelf, most of which is located in the Arctic. However, a lack of funds, technology, know-how and skills for the Arctic offshore operations forced Rosneft to sign strategic partnership agreements with foreign companies. By aggressively purchasing offshore licences, Rosneft managed to become the most significant player in the Russian offshore oil and gas sector. As of January 2017, it has held 56 licences for plots in the Arctic and in Russia's easternmost and southernmost seas (Table 2.2).

Table 2.2   Rosneft's offshore projects

Area	Sea	Projects
Western Arctic	Barents, Pechora and Kara Seas	19
Eastern Arctic	Laptev, East Siberian and Chukchi Seas	9
Far East	Okhotsk and Japanese Seas	20
Southern seas	Black, Azov and Caspian Seas	8

Source: Rosneft (2018a).

Having become the world's largest public oil company in terms of liquids production, Rosneft began to make a bet on new offshore projects with international partners (Table 2.3). In June 2010, Rosneft invited Chevron to participate in its Black Sea ventures: Val Shatskogo and the adjacent Tuapse Trough. Again, foreigners were required to finance the exploration phase (Mazneva and Novy 2010). However, in spring 2011, dissatisfied with the results of seismic exploration and unhappy with Rosneft's certain terms of cooperation, Chevron withdrew from the project (Melnikov 2011a).

Almost at the same time, on 27 January 2011, Rosneft and ExxonMobil reached an agreement on the joint development of hydrocarbon resources of the Black Sea – primarily the Tuapse Trough. Rosneft also decided to expand its offshore cooperation with BP. On 14 January 2011, the companies signed an agreement to develop the Arctic jointly and swap shares. However, despite the blessing of the then Prime Minister Vladimir Putin, the Russian oligarch shareholders of TNK-BP, known as the AAR consortium, torpedoed the

BP-Rosneft partnership (Poussenkova 2012). After the collapse of the deal with BP, Rosneft did not hesitate to find another partner. In summer 2011, Rosneft and ExxonMobil signed a strategic cooperation agreement.

In April 2012, following Gazprom's and Rosneft's strong lobbying, the government issued an ordinance aimed at providing fiscal and tariff incentives for the development of the Russian continental shelf (Rosneft 2012a). For example, offshore projects were given exemption from export duty as well as a lower rate of mineral production tax (5–15% depending on the complexity of the project); moreover, stability assurances of these terms were allowed for 15 years. Invited by Rosneft (and encouraged by the new fiscal benefits), other foreign companies – for example, ENI in April 2012 and Statoil (now Equinor) in May 2012 – gained a foothold in the Russian offshore sector. The terms for ENI, ExxonMobil and Equinor were almost identical: they were to bear the burden of the expenses at the exploration phase, and joint ventures were required to be established with Rosneft, which held 66.7% while the foreign company held 33.3% (Overland et al. 2013).

Table 2.3   *Rosneft's strategic partnerships*

Foreign companies	Main assets	Other partnership elements in Russia	International projects offered to Rosneft
ExxonMobil	Kara Sea (Vostochno-Prinovozemelskiye 1–3), Black Sea (Tuapse Trough)	ExxonMobil to assist Rosneft in developing hard-to-recover reserves in Western Siberia	Exploration projects in the Gulf of Mexico, the United States and Canada
ENI	Black Sea (Val Shatskogo), Barents Sea (Fedynsk and Tsentralno-Barentsevsk)	–	International projects primarily in North Africa
Equinor	Barents Sea (Perseevskiy), Okhotsk Sea (Magadan-1, Lisyanskiy, Kashevarovskiy)	Equinor to assist in the development of shale oil in the Samara region and the highly viscous oil of the Severo-Komsomolsk field in Yamal-Nenets Autonomous District	Norwegian offshore and international projects of Equinor

Source: Rosneft (2019a).

Rosneft also invited companies from other countries, such as PetroVietnam and Japanese INPEX, to cooperate on offshore developments, but these partnerships did not move beyond expressing good intentions. Because of issues related to both geology and international politics, cooperation with ENI, Equinor and ExxonMobil had limited success. In September 2016, Rosneft and Equinor drilled two exploration wells in the Okhotsk Sea at a depth of less than 150 metres, meaning that the Western sanctions would not apply. Yet both wells yielded only water (Fadeeva 2016b). In August 2014, Rosneft and ExxonMobil began drilling the northernmost well in Russia in the Kara Sea, the USD 600-million Universitetskaya-1. A month and a half later, partners discovered a new field; oil was found at the Vostochno-Prinovozemelskiy-1 block. However, due to the US sanctions, ExxonMobil had to halt its offshore operations in the Russian Arctic, which meant that Rosneft would not be able to proceed without its partner.

In April 2017, Rosneft began drilling the Central Olginskaya-1 well in the Laptev Sea, the first ever well in this sea. The Laptev Sea might contain 9.5 billion tonnes of oil equivalent; however, there are no ports in the vicinity of the well, and the navigation period is only two months per year. Drilling was being conducted by RN-Bureniye, the service division of Rosneft (Rosneft 2017b). This endeavour was Rosneft's acid test of whether it could conduct an Arctic offshore operation alone. In June 2017, Rosneft announced the discovery of a new field in this area (Markova 2017). However, it is still unclear whether the company will be able to develop this field on its own.

As the development of the Arctic is the key focus of Rosneft's offshore portfolio, the company has sought to build its capacities to manufacture offshore equipment and vessels capable of operating in the Arctic. Its main effort has been aimed at creating the Zvezda shipbuilding complex in the Russian Far East as part of a consortium involving Rosneftegaz and Gazprombank and creating partnerships with the leading international manufacturers of offshore equipment (Table 2.4). The several agreements that were signed after 2014 attest that sanctions did not entirely succeed in restricting Rosneft's access to international technology.

Table 2.4   Rosneft's main offshore technology partnerships

Foreign partner	Country	Year	(Possible) areas of cooperation
Siemens	Germany	2014	Shipbuilding, 'digital wharf', subsea systems, floating production storage and offloading (FPSO) units
General Electric	United States	2016	Manufacture of steering wheels for vessels, electronic naval equipment and dynamic positioning for ice-class vessels
Keppel/MH Wirth	Singapore	2016	Shipbuilding design and engineering
Hyundai Heavy Industries	Korea	2016	Aframax tankers
Samsung Heavy Industries	Korea	2018	Arctic shuttle tankers
Gaztransport & Technigaz	France	2017	LNG vessels

Shale Reserves

Rosneft is keen to develop unconventional reserves, but this also requires international partners. Hard-to-recover projects became commercially more attractive for foreign companies after amendments to the tax legislation introduced in September 2013. Under these amendments, the Abalac, Bazhenov, Domanic and Khadum formations were exempted from the mineral production tax for 15 years (Rosneft 2013d). Rosneft entered into partnerships to develop unconventional reserves with ExxonMobil, BP and Equinor (Rosneft 2013e, 2014).

However, Western sanctions created serious obstacles for these partnerships. For instance, the establishment of a joint venture between BP and Rosneft was frozen. The sanctions notwithstanding, the partners sometimes managed to bypass the restrictions: despite the EU sanctions, in early 2017, a joint venture between Rosneft and Equinor began drilling on the Domanic formations in the Samara region. Equinor seemingly managed to prove that these formations were limestone rather than shale and, therefore, not subject to the sanctions (Kozlov et al. 2017).

CSR

Rosneft publishes an annual sustainability report, covering issues related to labour protection, health, safety and environment (HSE) and the contributions of the company and its subsidiaries to the socio-economic development of the regions where they operate. The reports are produced in accordance with the requirements of the Global Reporting Initiative (GRI), the United Nations (UN) Global Compact and other nationally and internationally recognized sustainability reporting guidelines (see Rosneft 2017a, pp. 10–11 and Annex 2).

Gerhard Schroeder's opening message of the *2017 Sustainability Report* as chairman of the board hardly makes a reference to the environment, climate, society or sustainable development. Instead, it focuses on the sustained growth of the business. However, CEO Igor Sechin's message outlines Rosneft's strategic goals in this area, starting with HSE:

> Building a business that meets the highest HSE standards is an essential element of Rosneft's strategy. We are determined to make it to the upper quartile of global oil and gas companies in terms of HSE performance by 2022. (Rosneft 2017a, p. 5)

Sechin particularly emphasizes that Rosneft's membership of the UN Global Compact (since 2010) is a vehicle for promoting sustainability, 'focusing closely on innovations and initiatives to tackle climate change' (see section on climate change below). Sechin adds that 'innovation growth, biodiversity conservation and the sustainable use of water are also high on the priority list' (Rosneft 2017a, p. 5). In terms of social support, Sechin states thus:

> Rosneft is Russia's largest taxpayer and undertakes major social projects, many of which have completely changed the social landscape in both large cities and smaller communities by facilitating public access to modern medical, educational, recreational, sport and utility services. (Rosneft 2017a, p. 5)

In 2018, Gerhard Schroeder spoke about Rosneft and sustainable development and climate change:

> Rosneft is fully aware of the role it plays in advancing the sustainability agenda and is involved in efforts to address global issues. In a public statement released last year regarding its contribution to the 17 UN Sustainable Development Goals, the Company highlighted five priority goals on its agenda: Good Health and Well-Being, Affordable and Clean Energy, Decent Work and Economic Growth, Climate Action, and Partnerships for the Goals. (Rosneft 2018b, p. 3)

Sechin's opening message in the *2018 Sustainability Report* mentions HSE and the company's commitment to the UN Sustainable Development Goals:

> In the next four years, the company plans to join the first quartile of world oil and gas companies in terms of HSE. To strengthen the Company's position in the areas of environmental and social responsibility, Rosneft's ... commitment to the 17 UN Sustainable Development Goals was approved in 2018. (Rosneft 2018b, p. 5)

Rosneft has released joint declarations on environmentally sustainable Arctic offshore developments with its business partners ExxonMobil and Statoil (in 2012, now Equinor) and ENI (in 2013). In these agreements, the partners reaffirm their commitment to the sustainable development of the Arctic, including minimizing their impact on indigenous people and climate change (Rosneft 2012b). In 2014, Rosneft launched a marine conservation programme for the Arctic which included support for environmental protection, research and monitoring (Rosneft 2017a, p. 76). Rosneft's regional subsidiaries also fund scientific research programmes, such as Evenki reindeer protection in Krasnoyarsk region and taimen trout protection on Sakhalin Island.

In 2017, Rosneft updated its Sustainable Development Policy, with new sections on energy efficiency and conservation as well as emergency preparedness and response. This document also includes a human rights element, along with HSE and stakeholder engagement. It is publicly available online (Rosneft 2017c). Rosneft also has a Code of Business Ethics covering stakeholder and business relations and issues such as fraud, corruption and conflicts of interest. A growing area of interest and concern for Rosneft is cyber-security and the need to adapt effectively to the digital era.

Rosneft was ranked seventh in the 2018 World Wildlife Fund (WWF)/ Creon rating of the environmental performance of Russian oil and gas companies, which is a significant rise of ten places compared to 2017 (Shvarts et al. 2018). It achieved this through substantial improvements in its information disclosure, despite having dropped ten places in the ranking for its environmental impact.

In its sustainability report, Rosneft reports on its stakeholder engagement, noting that the most important types of engagement are public consultations on potential environmental impacts and regular roundtable meetings, which Rosneft has been holding since 2007 in host regions to share views and maintain partnership relations with local authorities and other stakeholders. In 2017, the company reported a total of 127 public awareness efforts of various types in different regions, including over 50 public consultations (for example, public hearings and surveys) to discuss offshore exploration activities and 15 roundtable meetings in various host regions (Rosneft 2017a, pp. 48–9).

In 2018, the company reported over 220 public awareness measures of various types in its regions of operation, including 16 public consultations and 16 roundtable meetings (Rosneft 2018b, p. 29).

Rosneft also has a strong focus on research and development (R&D), providing support to a range of specialized institutes and plans to establish a Technology Council to drive innovation throughout the company; it also has a programme for import substitution and localization (Rosneft 2017a).

In addition to its sponsorship of targeted R&D for its core business, Rosneft also supports the state education policy. Its representatives participate in the boards of trustees and the supervisory boards of nine higher education institutions. Rosneft helped establish a grammar school under the auspices of Moscow State University; it also finances the development of research and education infrastructure of naval higher education institutions, such as the Saint Petersburg State Naval Technical University and the Makarov State University of the Sea and River Fleet to train engineers for the shipbuilding industry. The company funds 17 chairs in the country's leading higher education institutions and provides some grants and scholarships to students.

Like other Russian oil companies, Rosneft signs agreements with the local authorities in the regions where it operates and funds individual charitable projects (in education and science, healthcare, sport, culture, 'revival of spiritual heritage' and environmental protection). For example, in September 2017, Rosneft and the Republic of Sakha-Yakutiya signed an agreement to finance social projects in the republic: Rosneft pledged to provide financial support for the construction of an education and laboratory building for the Yakutsk Academy of Sciences and a national research and education centre for 150 pupils specializing in physics, mathematics, sciences, information technology (IT) and engineering near Yakutsk (Rosneft 2017d).

As the national oil company (NOC) of Russia, Rosneft is tasked by the state with additional non-commercial functions, which transcend corporate social responsibility (CSR) in the way it is practised by international oil companies and, in return, receives additional benefits from the state. When Rosneft was weak, it welcomed the non-commercial functions, probably wishing to demonstrate its usefulness to its owner. However, as Rosneft grew stronger, its mindset changed and it began to present itself as a global energy company rather than a domestically oriented NOC; it also began to consider its social obligations as a burden that might prevent it from fulfilling its duty to its shareholders. Nonetheless, it believed that the additional benefits that accompanied the company for being an NOC belonged to it by right.

In the early 2000s, Rosneft was ordered to resuscitate Chechnya's oil sector, which had been devastated by war. This task combined CSR elements and support for Russian federalism and territorial integrity. In 2000, Rosneft established a subsidiary, Grozneftegaz (51% Rosneft; 49% government of Chechnya). Rosneft rebuilt 256 destroyed facilities and increased oil production in Chechnya from 0.7 million tonnes in 2001 to 2 million tonnes in 2004 (Neft i kapital 2004d, p. 37). However, this growth was not sustainable; by 2011, production had fallen back to 803 000 tonnes of oil and 0.3 billion cubic metres of gas, declining to 300 000 tonnes of oil in 2018. In 2010, Rosneft's board of directors voted to build a refinery in Grozny (capacity: 1 million tonnes per annum by 2013); however, the project stalled. Some progress was made in 2013 when Vladimir Putin ordered an evaluation of the possibility of building a refinery in Chechnya, and Rosneft was officially charged with the project. However, in 2016, it was postponed to a future unspecified date. The construction of a small refinery in Chechnya appeared unprofitable to Rosneft in the then macroeconomic environment, so it decided to forgo its non-commercial obligation in favour of its duty to its shareholders.

COPING WITH CHANGE

Predicting Oil Prices

Rosneft is fully aware of the importance of oil prices for its business. Thus, it highlighted in its *2016 Annual Report* that for its operational results the following main macroeconomic factors were of paramount importance: growth rate of the world economy and the Russian economy, the level of inflation in Russia and the rouble/dollar exchange rate and global oil and gas prices. Rosneft monitors the development of global oil prices continuously. The company's CEO is famous for his predictions (sometimes wrong, sometimes right) of the oil price. For example, in September 2014, when the oil prices had just begun falling, Igor Sechin said in an interview with TASS that 'The oil price will not drop below USD 90 per barrel. It is connected with several factors. But USD 90 is also a good price. It permits working. Rosneft's budget would not suffer. It is based on a price that does not exceed this level' (Igor Sechin, cited in Gordeyev 2017).

Sechin is particularly famous for his regular oil price predictions at the Saint Petersburg International Economic Forum, and the energy session where he usually delivers a presentation is one of the forum's most popular events. It is interesting to compare and verify the accuracy of his 2015, 2016 and 2017 predictions in Saint Petersburg.

In 2015, Sechin predicted that the oil price would grow to USD 170 by 2035. He also stated that the current oil price was characterized by 'elements of dumping' and the desire to redistribute the market; however, he believed that such phenomena could not be sustainable in the longer term. He stated that if the oil price did not correspond to objective requirements, both consumers and producers would suffer in the future (Igor Sechin, cited in Kalyukov 2015). Sechin argued that only minimal costs were factored into the current oil prices, which did not take into account the so-called full-cycle costs, in particular, the cost of decommissioning of depleted wells: 'The investments of transnational companies in major projects are already drastically decreasing. As a result, within 2–3 years, production might be significantly reduced' (Igor Sechin, cited in Kalyukov 2015). Igor Sechin also noted that many OPEC countries faced a severe budget deficit and tried to increase their revenues by exceeding the established production quotas: 'The policy of the USD 50–60 per barrel range has natural limitations in terms of time frame' (Igor Sechin, cited in Kalyukov 2015). He believed that the drastic drop of oil prices aggravated the fight for market shares, and, as a result, in May 2015, the imbalance between supply and demand in the oil market achieved a record high level of 3 million barrels per day: 'Some players make a bet on dumping, while others on fundamental, longer-term factors. Life will show who has been more successful and farsighted' (Igor Sechin, cited in Kalyukov 2015).

In 2016 at the Saint Petersburg Forum, Sechin noted the unpredictability of the main actors of the global oil market (Razumovskiy 2016). He warned that the balance which the market was just beginning to achieve could be disturbed at any moment and argued that Russia was one of the most reliable market actors: 'the market mechanisms of the functioning of the oil sector have been deformed' (Igor Sechin, cited in Razumovskiy 2016). He noted that there were many reasons for this situation: a focus on short-term financial instruments and the manipulation of market institutions to the detriment of the producer-consumer relations. While the market was moving towards a mid-term balance at the time, it might be disturbed at any moment because of the unpredictable behaviour of the producers who played the role of regulators in the global market (Razumovskiy 2016).

At the 2017 Saint Petersburg Forum, Igor Sechin said that low oil prices were here to stay. Although the OPEC+ agreement in late 2016 gave the market some breathing space, he believed its long-term effect would be limited. He reminded his audience that several major producers who did not join the agreement actively used the market situation to strengthen their own market positions, thus laying the ground for a new wave of instability. Shale oil producers in the United States were becoming major exporters, and production in Nigeria and Libya was growing. Investments in new projects began to decline and would affect the oil sector in the near future. Soon, the balance of the market at the level of USD 40 per barrel would cause about half of the world's producers to make a loss. He believed that Brazilian deep-water projects, Canadian oil sands and many US shale producers (except the most efficient sections of the Permian Basin) would suffer, while Russia, Saudi Arabia, Iran and several successful projects in the United States would remain competitive. He also stated that market uncertainty resulted in a growing fight over consumers and that all market participants were getting ready to raise their output (Newsru.com 2017).

In 2017, Igor Sechin made a prediction about the oil price in 2018. He believed that it would be in the range of USD 40–43 per barrel; in other words, it would return to the level of early 2016, mainly due to the growing production of shale oil (Gazeta.ru 2017). He was overly pessimistic – or perhaps he was just being a cautious, responsible manager – but he was right that shale oil would hold back the oil price. In 2018, Sechin predicted that the oil price in 2019 would be USD 53–60 per barrel, largely due to a rise in the interest rate in the United States (RBC.ru 2018).

Handling Oil Price Fluctuations and Sanctions

At the start of the 2008 financial crisis, Rosneft was the only company among the top five Russian oil companies that did not plan to cut back investments in 2009; instead, it decided to save on electricity and fuel and 'optimize' its staffing. It developed its 2009 budget realistically proceeding from the oil price of USD 50 per barrel (–26.5% compared to the 2008 level) (Malkova and Derbilova 2008).

In response to the 2008 crisis, Rosneft stopped recruiting new workers and sacked 9000 people, equivalent to some 5% of its staff. These cuts mainly affected temporary employees and workers close to the retirement age (Malkova 2009). During the first half of 2009, the company's operating expenses (including salaries) fell by 12.3% to USD 1.8 billion, while administrative and general economic expenses (including salaries of personnel at headquarters and management of subsidiaries) fell by 14.3% to USD 676 million. On the other hand, it paid outsize bonuses in 2008 to personnel at headquarters where 1537 people worked.

In the years after 2008, Rosneft continued to compensate for falling oil prices by raising efficiency and exercising strict cost control. Some analysts believed that it was the most efficient company in the Russian petroleum sector in that period (Malkova 2010).

Another measure that aimed at cushioning the impact of low oil prices involved lobbying for tax benefits for Vankor, which yielded successful results. Tax breaks included reducing the export duty to some 30% of the standard rate. This helped the company save up to USD 3 billion in 2010 (Mazneva et al. 2010) – a privilege that would be revoked in 2011. However, Rosneft requested the government to prolong the partial tax holiday for Vankor until 2014 when the field was expected to reach its peak production of 25 million tonnes per annum. In return, the company guaranteed the Ministry of Finance a fixed amount of fiscal revenues (Gavshina 2011).

Rosneft had to respond to the 2014 oil price collapse, the economic crisis in Russia and Western sanctions concurrently. Since these developments, alongside the colossal debts incurred in connection with numerous acquisitions from 2004 onwards, seriously strained the company's financial position, it used its considerable lobbying clout to obtain financial benefits from the government and more loans from China.

Due to the sanctions, Rosneft lost about USD 20 billion in capitalization by mid-2015. The losses were genuine because credits and equipment became more expensive, and many opportunities were missed because the Arctic offshore projects were halted. In early 2014, before the events in Ukraine, Rosneft's value on the London Stock Exchange (LSE) was some USD 80 billion; by mid-2015, it was about USD 40 billion, although this decline also partly resulted from low oil prices.

Continuing this trend, Rosneft lobbied for unprecedented tax benefits for the Samotlor field in October 2017. The mineral production tax was reduced by RUB 35 billion for ten years (although, initially, Rosneft had sought a reduction of RUB 70 billion); in return, Rosneft promised to increase investments in the field and raise production (Kozlov and Barsukov 2017). In 2016, investments in Samotlor grew by 38%, although production declined moderately.

In 2014, as an anti-crisis measure, Rosneft attempted to obtain money from the National Wealth Fund for its projects. The National Wealth Fund is Russia's stabilization fund and, according to the Ministry of Finance, is intended to help ensure the long-term stability of Russian pensions. In 2014, Rosneft applied to the National Wealth Fund for RUB 2.4 trillion to aid the development of East Siberian gas fields and the construction of the Eastern Petrochemical Company (Papchenkova and Tretyakov 2014). In January 2015, Rosneft increased the number of projects in the application for financing from the National Wealth

Fund from 12 to 28 (Papchenkova et al. 2015). The Ministry of Economic Development proposed to limit the request to five top priority projects (Zvezda, the gas fields of Rospan, Russkoye and Yurubcheno-Tokhomskoye fields and the Achinsk refinery). The Ministry of Finance was, in principle, against the financing of oil projects from the National Wealth Fund and tried to bury Rosneft's application in endless delays (Papchenkova and Starinskaya 2015). Finally, in July 2015, Vladimir Putin decided not to allocate that money to Rosneft (Papchenkova and Fadeeva 2015).

The sanctions and low oil prices also forced Rosneft to revise its plans for offshore exploration. Rosneft requested Rosnedra to freeze the licences of certain projects in the Barents, Okhotsk, East Siberian and Pechora Seas in 2015 and the Black Sea in 2017 mainly because of the adverse effects of the Western sanctions, including the loss of foreign partners; these requests were partially granted (Kommersant 2017; Starinskaya 2015).

In short, Rosneft's main response to low prices, crises and sanctions has involved attempts (successful and unsuccessful) to halt certain projects and the use of its political connections to obtain financial benefits and financial support from the state. However, this point should not be overstated. Novatek, which is a fully privately owned company, received the funds it requested from the National Wealth Fund. Rosneft ultimately ended up relying more on Chinese capital.

Climate Change

As the world's largest publicly traded oil company by production, Rosneft's activities inevitably have a significant impact on the climate. However, Rosneft's stance on climate change is profoundly ambivalent.

According to its *2017 Sustainability Report*, Rosneft sets GHG emissions reduction targets. The sustainability report also details the company's emissions for that year and its ongoing efforts to reduce them, including its Gas Investment Programme and Energy Efficiency Programme with a strong focus on the utilization of APG. In 2017, Rosneft's GHG emissions amounted to 76 million tonnes of CO_2 equivalent, including direct emissions of 54 million tonnes and indirect emissions of 22 million tonnes related to heat and power consumption (Rosneft 2017, p. 82). This represented a 2% decrease from the previous year despite the acquisition of Bashneft. The *2018 Sustainability Report* states that the GHG emissions amounted to 76.4 million tonnes of CO^2 equivalent, including 54.2 million tonnes of direct emissions and 22.2 million tonnes of indirect emissions. This means that they rose slightly from the preceding year (Rosneft 2018b, p. 73).

According to the Carbon Disclosure Project's 2017 *Carbon Majors Report*, the majority of the world's carbon emissions are the responsibility of 100 companies, many of which (unsurprisingly) are oil and gas companies (CDP 2017). Rosneft came 27th in this list, with lower cumulative emissions than Gazprom, ExxonMobil, Shell, BP, Total, ConocoPhillips and LUKOIL. According to their methodology, Rosneft's cumulative emissions between 1988 and 2015 represented 0.7% of global emissions. Gazprom came third with 3.9% of global emissions.

In its *2016 Sustainability Report*, Rosneft claims that it is one of the pioneering oil companies in the world in terms of GHG emissions control. It says that it was the second-lowest among its competitors in terms of the ratio of the total volume of GHG emissions per thousand barrels of oil equivalent of production in 2015, according to Bloomberg's methodology. Rosneft also states that it had the lowest ratio of energy consumption per thousand barrels of oil equivalent of production in 2015 among peer companies (Rosneft 2016f, p. 29). According to the *2017 Sustainability Report*, Rosneft has also been active in helping the government to develop effective climate legislation 'to enhance government regulation mechanisms for GHG emissions and drive the ratification procedure for the Paris Agreement', including efforts to reduce emissions in the public sector (Rosneft 2017a, p. 75).

In its *2018 Sustainability Report*, Rosneft noted several measures to reduce GHG emissions such as reduced business trips, and increased emphasis on video conferencing. Moreover, the company launched several environmentally friendly gas filling stations and low emission products (new Pulsar 100 gasoline, Euro 6 and BP Active Fuel) (Rosneft 2018b, p. 76).

In the past, Rosneft was one of the worst offenders in Russia in terms of flaring APG; however, it is now making efforts to improve its performance in this area. For instance, in 2010, it commissioned a gas-processing facility and a 180 MW gas turbine power station fuelled with APG in the Priobskoye field (Rosneft 2011). In 2011, Rosneft increased the level of APG utilization at Sevmorneftegeofizika to 95% within the framework of its Target Gas Programme. Rosneft reports that due to the implementation of its Gas Investment Programme and Energy Efficiency Programme, APG flaring from 2013 to 2016 decreased from 29% to 9% (Rosneft 2016f, p. 161). In 2018, Rosneft's level of APG utilization was 84%, that is, lower than in 2017 (89%) (Rosneft 2019b).

Rosneft has been involved in several joint implementation projects under the Kyoto Protocol. For instance, in March 2012, the Ministry of Economic Development approved Rosneft's APG utilization investment projects as a joint implementation project (Rosneft 2012d). The decision of the Ministry permitted Rosneft to sell units of GHG emission reductions to the investment company Carbon Trade & Finance SICAR.

Although it may appear that Rosneft has been proactive about climate change and reported on its progress in its *2017* and *2018 Sustainability*

Reports, this has not always been the case, and improvements in reporting are the result of pressure from investors. It seems that Rosneft tends not to perceive all these issues as low-carbon alternatives aimed at climate change mitigation but rather as a chance to earn more money. Igor Sechin's ambivalent attitude can probably explain this indifference to climate change since Rosneft is very much a one-man show. According to Armitage (2015), Sechin once joked about how he had responded to a question from the audience about renewable energy at a public event:

> I can tell you I am not a big expert in that area, but I know a few things. First of all, we are the subject of global climate change cycles. I actually comforted the guy. I said, 'those cycles repeat every 30 million years, so everything is normal. The human effect on the environment is less than any volcano. A volcanic eruption produces more CO_2 than any human activity. The rotting of algae in the ocean significantly exceeds any human-made effect, so one should be calm about it'. (Igor Sechin, cited in Armitage 2015)

At the 2017 Saint Petersburg Forum, Igor Sechin discussed the prospects of electric cars, believing that they would occupy a market niche but not at the level anticipated by some actors. In general, he concluded that 'hydrocarbons were and would be in demand' (Igor Sechin, cited in Yeremenko 2017).

Internationalization

Rosneft's long-term objective is to become a global energy company. The company is internationalizing in two main directions: entering the upstream operations of other oil-producing countries around the world and building niches in the European and Asian downstream markets (Tables 2.5 and 2.6). Through its international partners in the Russian petroleum sector, Rosneft can have access to projects in third countries where the partners are well respected. It also interacts with other national oil companies overseas, both upstream and downstream.

Rosneft's international expansion has multiple drivers. First, like other major oil companies, Rosneft aims to diffuse risks and grasp different markets. Second, it hopes it can acquire technological know-how from foreign partners. Third, the company acts as a petroleum ambassador for Russia, helping to establish relations or strengthen ties with strategic allies, such as China, Cuba and Venezuela. Despite this, Rosneft's close links to the Russian state are sometimes a disadvantage since it is regarded in many countries as a tool of the Kremlin. Finally, like other major Russian companies, Rosneft's international forays reflect a unique post-Soviet yearning to be seen as an equal on the international scene with its unique agency and not as an entity that is mistrusted.

Table 2.5   Rosneft's upstream international projects

Country	Time	Areas/assets	Partners	Comments
Kazakhstan	Early 2000s–	Adaisk Zone and Kurmangazy structure	Sinopec, Kazmunaigas	Unsuccessful (Kurmangazy-Kazakhstan Rosneft 2018)
Algeria	2001–17	Block 245, Southern Illizi Province	Sonatrach, Stroytransgaz	Discovered three fields; Rosneft lost faith in the project in 2017 (Kozlov 2017)
UAE	2010–16	Sharjah	Crescent Petroleum	(Interfax 2017)
Egypt	2017	Zohr field	ENI, EGAS	(Rosneft 2016b)
Mozambique	2014–	Angoche River Basin, Zambezi Delta	ExxonMobil	Sanctions hinder cooperation in Russia; better prospects abroad
Venezuela	2008–	Junin-6, Carabobo-2, Patao, Mejillones, Rio Caribe	Other Russian oil companies, PDVSA	Venezuela is Russia's main political ally in the Americas. Attractive assets were acquired along with TNK-BP (Melnikov 2011; Rosneft 2015)
Cuba	2014–	Varadero-Central Block	RN-Exploration, CUPET	
Brazil	2014–	Solimoes	HRT, Petrobras	Acquired along with TNK-BP
USA	2013	Mexican Gulf	ExxonMobil	(Interfax 2017)
Canada	2012	Cardium Project		Strongly affected by sanctions (Interfax 2017)
Norway	2013	Twenty-second licensing round	Equinor	
Vietnam	2010	Nam Con Son	PetroVietnam	Acquired along with TNK-BP

Table 2.6   *Rosneft's downstream international projects*

Country	Time	Areas/assets	Comments
Kyrgyzstan	2014	Bishkek Oil Company	
Ukraine		Lisichansk refinery	
Belarus		Mozyr refinery	Acquired along with TNK-BP
Germany	2011–	Stakes in MiRO, PCK, Schwedt and Bayernoil refineries	In brief, Rosneft became the third-largest refiner in Germany
Italy	2013–17	Saras	Hampered by international sanctions against Russia (Kezik 2013)
India	2017–	Vadinar refinery	

Despite its ambitious globalization plans and large number of international projects, almost all of Rosneft's actual output is still concentrated in Russia, and the share of overseas projects in its oil production portfolio is only about 1% (Interfax 2017). According to Chris Inchcombe, the Director of the International Projects Department, 'a company such as Rosneft should not strive to get access to all the regions of the world. We'll focus on where we already work and where we see real opportunities' (Interfax 2017).

Historically, Russia has always focused on energy exports to Europe. However, during the 2000s, Russia's attention started shifting from Europe, where demand was stagnating, to the Asia-Pacific market, especially China, while strengthening political and economic ties with both China and India. Another driver was the national interest in revitalizing Eastern Siberia and the Far East regions, which constitute 60% of Russia's territory but account for only 10% of its population. It was also necessary to establish a new petroleum province to supplement the ageing Western Siberian fields that currently account for 60% of Russia's crude production; Eastern Siberia and the Far East were next in line.

Thus, during the 2000s, Rosneft gradually displaced YUKOS and TNK-BP in the strategic Eastern region and began to implement the Asian vector of state energy policy. In addition to the national interests noted above, Rosneft also had its own long-term strategic considerations for going east: it wanted to carve out a solid chunk of the rapidly growing Chinese market and diversify its export routes. Meanwhile, close cooperation with Chinese NOCs also facilitated Rosneft's handling of the massive debts it had incurred through its policy of aggressive acquisition.

There was a breakthrough in Russian-Chinese oil relations when Rosneft acquired Yuganskneftegaz in 2004 and the Chinese banks lent Rosneft USD 6 billion to repay its debts (see above). Later, Rosneft actively lobbied for the construction of the ESPO pipeline and the spur to China. Rosneft is the main (and actively growing) supplier of Russian oil to the Asia-Pacific region (Overland and Kubayeva 2018) (Figure 2.1, see p. 50).

The share of Russian oil in the Chinese market is steadily growing. In 2014, Russia became the third supplier to China after Saudi Arabia and Angola. In 2016, the largest portion of China's 7.6 million barrels per day of crude imports was supplied by Russia (14% of total imports), followed by Saudi Arabia (13%), Angola (11%) and Iraq (10%) (Barron 2017). Sanctions became an additional driver of Rosneft's expansion to the Asia-Pacific, and in 2017, Russia delivered 1.2 million barrels per day to China, 14% higher than in 2016 and ensuring that it remained the number one external supplier of oil to China (Vestifinance 2018). In 2018, Russia remained the number one external supplier of oil to China, with 1.43 million barrels per day (Oilcapital.ru 2019).

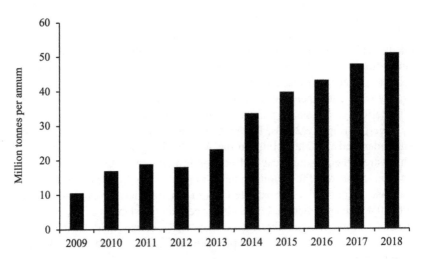

Figure 2.1   *Rosneft's oil exports to the Asia-Pacific region (2009–18), million tonnes per year*

Source: Rosneft's annual reports.

At the same time, since the mid-2000s, thanks to Rosneft, Chinese companies have also had access to the Russian upstream sector. In 2005, Rosneft invited Sinopec to buy 25% of the Veninsk block of Sakhalin-3. Sinopec committed itself to cover a proportion of Rosneft's exploration expenses and provide a certain amount of financing at the development phase (Rosneft 2012c). In August 2006, Sinopec purchased from TNK-BP 96.96% of its subsidiary Udmurtneft for USD 3.5 billion through the company Promleasing. In December 2006, Rosneft exercised an option to buy 51% of Promleasing from Sinopec (AK&M 2006).

Although the Chinese are highly interested in equity oil, they do not want to overpay for assets. As a result, certain Rosneft plans to establish E&P partnerships with Chinese NOCs have failed. For instance, in 2013, Rosneft (51%) and CNPC (49%) agreed to form a joint venture based on Taas-Yuryakh Neftegazodobycha; however, the deal fell apart because the partners could not agree on the terms. In the same year, Rosneft signed several agreements with CNPC on joint activities in the Barents and Pechora Seas; however, disagreements over the prices, high costs and unclear economics of Arctic exploration hampered progress.

Also, on 9 November 2014, Rosneft and CNPC signed a framework agreement on the purchase of 10% of Vankorneft. The Chinese were apparently ready to buy a larger stake of Vankorneft; however, the deal fell through because the partners could not agree on the price. Later, on 7 November 2016, Rosneft and Beijing Gas Group signed an agreement to sell 20% of the shares of Verkhnechonskneftegaz for some USD 1.1 billion; they also reached an agreement on cooperation in the gas sector.

Since relations with China were progressing mainly in oil-for-loans deals and failing in the joint development of Russian resources, Rosneft decided to renew its cooperation with India, which had initially been launched in 2001, by selling half of its share in Sakhalin-1 to the Indian Oil and Natural Gas Corporation (ONGC). In May 2016, Rosneft and ONGC closed the deal on the sale of a 15% stake in Vankorneft for USD 1.27 billion. In October 2016, Rosneft sold another 11% of Vankorneft for USD 930 million to ONGC, making it the owner of 26% of Rosneft's subsidiary (Rosneft 2016c). Another deal was closed on 5 October 2016 for the purchase of 24% of Vankorneft by a consortium of Indian investors (Oil India, Indian Oil and Bharat PetroResources) for USD 2 billion (Rosneft 2016d). Thus, the stake of Indian companies in Vankorneft grew to 49.9%. Rosneft retained a 51.1% majority in the board of directors; it also maintained its control over Vankorneft's operations and the production infrastructure of the Vankor cluster.

Vankor became the most significant, albeit not the only, breakthrough of India in the Russian upstream petroleum sector. In October 2016, Rosneft, Oil India, Indian Oil and Bharat PetroResources also agreed to trade 30% of Taas-Yuryakh for USD 1.2 billion (Rosneft 2016e).

Rosneft also secured a niche in the downstream sector in India, the success of which should particularly be attributed to its ties with the Indian company Essar (Rosneft 2016a, p. 105). In August 2017, Rosneft and a consortium of investors closed a deal on buying 49% of Essar Oil Limited (EOL) for USD 13 billion. Vadinar refinery, with a throughput capacity of 20 million tonnes per annum and a conversion ratio of 95.5%, forms part of EOL. EOL also owns a network of 2.7 thousand fuel stations in India (Rosneft 2017e).

Under the pressure of sanctions, Rosneft managed to find another Asian partner. In October 2016, Rosneft and the Indonesian company Pertamina signed preliminary agreements envisaging the sale of stakes in the Russkoye field (up to 37.5%) and in the northern part of Chaivo (up to 20%). The partners also established a joint venture – Rosneft (45%) and Pertamina (55%) – to build the 15-million-tonne-per-annum Tuban refining and petrochemical complex on the island of Java (Fadeeva 2016c).

CONCLUSION

Rosneft mirrors the changes that took place in Russia between 1991 and 2020 as well as those that it sometimes initiated itself. As a response mechanism to external changes, Rosneft has often drawn on its formidable lobbying potential to protect itself or benefit from such changes, albeit with varying degrees of success. At critical times, Rosneft has been able to make use of state funds and state-sanctioned international relations to secure deals and develop its business. In return, it has provided support to the state in critical areas, such as development in post-war Chechnya or the less developed eastern regions of Russia; it has also played a crucial role in international energy diplomacy, notably with China. With Igor Sechin as CEO, Rosneft has launched many upstream and downstream international projects, driven by political and/or commercial reasons.

The company has managed to handle the challenges of fluctuating oil prices and the Western sanctions, partly because of its consistent monitoring of the oil prices and partly because of its close ties with the Kremlin; its success can also be ascribed to various strategic partnerships with foreign oil companies. Rosneft has also benefited from its diverse portfolio of projects within Russia thanks to numerous strategic acquisitions, state influence and access to financing. The company owns many downstream assets in both Russia and Europe, ranging from refineries to petrochemical businesses, gas-processing plants and gas stations.

Among its peers, Rosneft can hardly be distinguished for its environmental performance and standard social projects. However, the company appears to have modified its policy on transparency and reporting, notably concerning reporting on its GHG emissions and efforts to reduce its climate change impact. Having been one of the worst offenders among its peers in the past, Rosneft has been making concerted efforts to reduce its APG flaring. The company came out relatively favourably in a major report on the climate responsibility of major companies, although it remains among the top 30 international companies responsible for the largest part of global climate change, according to the Carbon Disclosure Project.

NOTES

1. This chapter title is partly inspired by Poussenkova (2007).
2. 'On Specifics of Privatization and Corporatization of the State Enterprises, Production and R&D Associations of Oil Production, Refining and Marketing Segments'.
3. 'On Top Priority Measures to Improve Activities of the Oil Companies'.
4. The Samotlor field produced 150 million tonnes per annum at the peak in the 1980s. However, its production in 2019 was only some 20 million tonnes per annum. Due to excessive flooding during the communist era, the water-cut at Samotlor is 95.8%.

REFERENCES

AK&M (2006), 'Rosneft planiruyet realizovat optsion na pokupku 51% aktsiy Udmurtnefti', accessed 15 October 2018 at https://www.vedomosti.ru/library/news/2006/06/20/rosneft-planiruet-realizovat-opcion-na-pokupku-51-akcij-udmurtnefti.

Armitage, J. (2015), 'Igor Sechin: The oil man at the heart of Putin's Kremlin', *Independent*, accessed 12 October 2018 at http://www.independent.co.uk/news/business/analysis-and-features/igor-sechin-the-oil-man-at-the-heart-of-putins-kremlin-10043230.html.

Barron, J. (2017), 'More Chinese crude oil imports coming from non-OPEC countries', accessed 12 October 2018 at https://www.eia.gov/todayinenergy/detail.php?id=30792.

Barsukov, Y., D. Butrin and D. Kozlov (2016), 'Ne mytiem, tak Katarom', accessed 12 October 2018 at https://www.kommersant.ru/doc/3164301.

Bloomberg (2019), 'ROSN: LI', accessed 3 January 2019 at https://www.bloomberg.com/quote/ROSN:LI.

CDP (2017), The Carbon Majors Database, accessed 13 October 2019 at https://6fefcbb86e61af1b2fc4-c70d8ead6ced550b4d987d7c03fcdd1d.ssl.cf3.rackcdn.com/cms/reports/documents/000/002/327/original/Carbon-Majors-Report-2017.pdf?1501833772.

Derbilova, Y. (2006), 'Vsya pravda Rosnefti', accessed 11 October 2018 at https://www.vedomosti.ru/newspaper/articles/2006/05/17/vsya-pravda-rosnefti.

Derbilova, Y. and V. Kudinov (2005), 'Bez Sberbanka ne oboshlis', accessed 11 October 2018 at https://www.vedomosti.ru/newspaper/articles/2005/08/26/bez-sberbanka-ne-oboshlis.

Derbilova, Y. and I. Reznik. (2010), 'Sergey Bogdanchikov pokinet Rosneft', accessed 12 October 2018 at https://www.vedomosti.ru/newspaper/articles/2010/09/01/dolgie_provody.

Derbilova, Y. and V. Surzhenko (2006), 'Rosneft zadachu vypolnila', accessed 12 October 2018 at https://www.vedomosti.ru/newspaper/articles/2006/07/17/rosneft-zadachu-vypolnila.

Derbilova, Y. and A. Tutushkin (2004), 'Nakhodchivaya Rosneft', accessed 23 January 2019 at https://www.vedomosti.ru/newspaper/articles/2004/03/24/nahodchivaya-rosneft.

Fadeeva, A. (2016a), 'Igor Sechin: Rosneft nuzhno prodavat pri tsene nefti v $100 za barrel', accessed 12 October 2018 at https://www.vedomosti.ru/business/articles/2016/02/11/628482-sechin-rosneft-nuzhno-prodavat-tsene-nefti-100-barrel.

Fadeeva, A. (2016b), 'Rosneft i Statoil nashli vodu v Okhotskom more', *Vedomosti*, accessed 12 October 2018 at https://www.vedomosti.ru/business/articles/2016/09/29/658943-rosneft-statoil.

Fadeeva, A. (2016c), 'Rosneft prodaet indonezioskoi Pertamina doli v dvuh mestorozh-deniyah', accessed 12 October 2018 at https://www.vedomosti.ru/business/articles/2016/10/06/659794-rosneft-pertamina-mestorozhdeniyah.

Fadeeva, A. (2016d), 'Na Vankore nachalos snizheniye dobychi', accessed 23 January 2019 at https://www.vedomosti.ru/business/articles/2016/06/09/644700-vankore-dobichi.

Fadeeva, A., M. Papchenkova and T. Voronova (2016a), 'Rosneft doplatit budzhetu za sobstvennye aktsii', accessed 12 October 2018 at https://www.vedomosti.ru/business/articles/2016/12/12/669144-rosneft-doplatit.

Fadeeva, A., Y. Derbilova, M. Papchenkova and A. Terentyeva (2016b), 'Bashneft privatizirovana', accessed 15 October 2018 at https://www.vedomosti.ru/business/articles/2016/10/13/660741-bashneft-privatizirovana.

Fjaertoft, D. and U. Overland (2015), 'Financial sanctions impact Russian oil, equipment export ban's effects limited', *Oil and Gas Journal*, **8** (113), 66–72.

Gavshina, O. (2011), 'Rosneft opyat lishilas britanskogo partnera', *Vedomosti*, accessed 12 October 2018 at https://www.vedomosti.ru/business/articles/2011/12/27/vr_sdala _sahalin.

Gazeta.ru (2017), 'Tseny na neft: Vperedi padeniye', accessed 12 October 2018 at https://www.gazeta.ru/business/2017/09/11/10882958.shtml.

Gordeyev, V. (2017), 'Sechin predskazal vozvrasheniye tsen na neft k urovnyu 2016 goda', *RBC*, accessed 12 October 2018 at https://www.rbc.ru/finances/11/09/2017/59b617009a794709393eb56e.

Hudson, J. and N. Poussenkova (1996), *Russian Oil, Prospects for Progress: Industry Background and Status*, Volume 4 of Report on Russian Oil, London: Salomon Brothers.

Interfax (2017), 'Kris Inchkomb: Rosneft ne stremitsya dobyvat vo vsekh regionakh mira', accessed 12 October 2018 at https://www.interfax.ru/interview/550161.

Kalyukov, Y. (2015), 'Sechin predskazal rost neftyanykh tsen do $170 za barrel k 2035 godu', *RBC*, accessed 12 October 2018 at https://www.rbc.ru/economics/19/06/2015/55843a319a79477a7a20e21f.

Kezik, I. (2013), 'Rosneft pokupayet dolyu v italianskom NPZ', accessed 12 October 2018 at https://www.vedomosti.ru/business/articles/2013/04/16/rosneft_propisalas _na_sardinii.

Kommersant (2017), 'Rosneft iz-za sanktsiy priostanovila dobychu nefti na uchastke v Chernom more', accessed 12 October 2018 at https://www.kommersant.ru/doc/3453648.

Kozlov, D. (2017), 'Rosneft ishet vyhod iz Alzhira', accessed 12 October 2018 at https://www.kommersant.ru/doc/3186781.

Kozlov, D. and Y. Barsukov (2017), 'Rosneft doburila do lgot', accessed 12 October 2018 at https://www.kommersant.ru/doc/3433278.

Kozlov, D., A. Zanina and Y. Khvostik (2017), 'Statoil probilas skvoz slantsy', accessed 12 October 2018 at https://www.kommersant.ru/doc/3206767.

Kurmangazy-Kazakhstan Rosneft (2018), 'Istoriya', accessed 12 October 2018 at https://kurmankazah.rosneft.ru/about/Glance/OperationalStructure/Razvedka/kurmankazah/.

Latyshova, E. (1995), 'Purneftegas uverenno shagaet vzad-vpered', *Neft i kapital*, **3**, 46–51.

Malkova, I. (2009), 'V Rosnefti i TNK-VP massovye uvolneniya', accessed 12 October 2018 at https://www.vedomosti.ru/business/articles/2009/10/02/v-rosnefti-i-tnk-bp -massovye-uvolneniya.

Malkova, I. (2010), 'Rosneft v krizisnyy god uluchshila rentabelnost po EBITDA', accessed 12 October 2018 at https://www.vedomosti.ru/business/articles/2010/02/02/rosneft-v-krizisnyj-god-uluchshila-rentabelnost-po-ebitda.

Malkova, I. and Y. Derbilova (2008), 'Naperekor krizisu', accessed 12 October 2018 at https://www.vedomosti.ru/newspaper/articles/2008/11/10/naperekor-krizisu.

Markova, N. (2017), 'Gosorgany podtverdili otkrytiye Rosneftyu Tsentralno-Olginskogo mestorozhdenya na Taymyrye', accessed 12 October 2018 at http://neftianka.ru/gosorgany-podtverdili-otkrytie-rosneftyu-centralno-olginskogo-mestorozhdeniya -na-tajmyre/.

Mazneva, Y. and Y. Derbilova (2007), 'Rosneft spryatalas', accessed 11 October 2018 at https://www.vedomosti.ru/newspaper/articles/2007/05/18/rosneft-spryatalas.

Mazneva, Y. and V. Novyy (2010), 'Rosneft i Chevron sobirayutsya potratit na Chernom more 1 trln rub', accessed 12 October 2018 at https://www.vedomosti.ru/business/articles/2010/06/18/rosneft-i-chevron-sobirayutsya-potratit-na-chernom-more-1-trln-rub.

Mazneva, Y., P. Sterkin and A. Fialko (2010), 'Rosneft nashla soyuznika na 3 milliarda', accessed 12 October 2018 at https://www.vedomosti.ru/newspaper/articles/2010/09/30/rosneft_nashla_soyuznika_na_3_mlrd.

Melnikov, K. (2011a), 'Rosneft zaderzhitsya na Chernom more', accessed 12 October 2018 at https://www.kommersant.ru/doc/1621563?query= роснефть%20черное%20море.

Melnikov, K. (2011b), 'Rosneft podderzhit Venezuelu', accessed 12 October 2018 at https://www.kommersant.ru/doc/1827175.

Mescherin, A. (2005), 'Riskovannaya klounada', *Neftegazovaya vertikal*, 1, 5.

Neft i kapital (2003), 'Prirazlomnoe mestorozhdeniye: Vse uchteno', accessed 12 October 2018 at https://neftegaz.ru/news/view/42801-Prirazlomnoe-mestorozhdenie-vsyo-uchteno.

Neft i kapital (2004a), 'Rosneft', *Neft i kapital*, 10, 14.

Neft i kapital (2004b), 'Purneftegaz', *Neft i kapital*, 10, 60.

Neft i kapital (2004c), 'Sergei Bogdanchikov', *Neft i kapital*, 10, 161.

Neft i kapital (2004d), 'Neftyaniki vosstanavlivayut dobychu chernogo zolota v Chechne', accessed 12 October 2018 at https://neftegaz.ru/news/view/44660-Neftyaniki-vosstanavlivayut-dobychu-chernogo-zolota-v-Chechne.

Newsru.com (2014), 'Sechin: Sanktsii protiv Rosnefti neobosnovannye, subyektivnye i nezakonnye', accessed 12 October 2018 at https://www.newsru.com/finance/17jul2014/setchinsanct.html.

Newsru.com (2017), 'Sechin: nizkiye tseny na neft – eto nadolgo', accessed 12 October 2018 at https://www.newsru.com/finance/02jun2017/sechinforum.html.

NGFR (2008), 'Rosneft', accessed 23 January 2018 at http://www.ngfr.ru/library.html?rosneft.

Oilcapital.ru (2019), 'Eksport rosiyskoi nefti v KNR v dekabre dostig 7.04 mln tonn', accessed 26 November 2019 at https://oilcapital.ru/news/markets/25-01-2019/rossiya-snova-na-pervom-meste-po-postavkam-v-kitay.

Overland, I. and G. Kubayeva (2018), 'Did China bankroll Russia's annexation of Crimea? The role of Sino-Russian energy relations', in Helg Blakkisrud and Elana Wilson Rowe (eds), *Russia's Turn to the East: Domestic Policymaking and Regional Cooperation*, Cham: Palgrave Macmillan, pp. 95–118.

Overland, I., J. Godzimirski, L.P. Lunden and D. Fjaertoft (2013), 'Rosneft's offshore partnerships: The re-opening of the Russian petroleum frontier?', *Polar Record*, 249 (49), 140–53.

Papchenkova, M. and A. Fadeeva (2015), 'Vladimir Putin ne dast deneg Rosnefti', accessed 12 October 2018 at https://www.vedomosti.ru/business/articles/2015/08/06/603694-rosnefti.

Papchenkova, M. and G. Starinskaya (2015), 'Minfin ne khochet finansirovat neftyanye proyekty', accessed 12 October 2018 at https://www.vedomosti.ru/business/articles/2015/06/03/594925-minfin-ne-hochet-finansirovat-neftyanie-proekti.

Papchenkova, M. and P. Tretyakov (2014), 'Rosneft mozhet ne dozhdatsya iskomykh 2.4 trln rubley', accessed 12 October 2018 at https://www.vedomosti.ru/business/articles/2014/11/06/vtoroj-krug-rosnefti.

Papchenkova, M., M. Lyutova and P. Tretyakov (2015), 'Rosneft uvelichila spisok proektov v zayavke na dengi iz FNB', accessed 12 October 2018 at https://www .vedomosti.ru/business/articles/2015/01/23/rosnefti-malo-ne-byvaet.

Petleva, V. (2018), 'Uiti po kitaiski', accessed 23 January 2019 at https://www .vedomosti.ru/opinion/articles/2018/05/08/768869-uiti-po-kitaiski.

Petrachkova, A. and E. Derbilova (2007), 'Ne podelili Tomskneft', accessed 15 October 2018 at https://www.vedomosti.ru/newspaper/articles/2007/07/13/ne -podelili-tomskneft.

Politinformatsia (2015), 'Polnye sanktsionnye spiski SSHA, ES i Kanady protiv rossiskikh grazhdan i kompaniy', accessed 12 October 2018 at http://politinform .su/pervaya-polosa/21275-polnye-sankcionnye-spiski-ssha-es-i-kanady-protiv -rossiyskih-grazhdan-i-kompaniy.html.

Poluektov, N. (1998), 'Bazhaev vozvraschaetsya v Rosneft', accessed 23 January 2019 at https://www.kommersant.ru/doc/209543.

Poussenkova, N. (2007), *Lord of the Rigs: Rosneft as a Mirror of Russia's Evolution*, Houston, TX: James Baker Institute Publications.

Poussenkova, N. (2012), *Novye Zvezdy Mirovoi Neftyanki: Istoriya Uspekhov i Provalov Natsionalnykh Neftyanykh Kompaniy*, Moscow: Ideya Press.

Poussenkova, N. and I. Overland (2018), 'Russia: Public debate and the petroleum sector', in Indra Overland (ed.), *Public Brainpower: Civil Society and Natural Resource Management*, Cham: Palgrave, pp. 261–89, accessed 3 January 2019 at https://www.researchgate.net/publication/320657842.

Razumovskiy, P. (2016), 'Rossiyskiy otvet na neftyanye potreseniya', accessed 12 October 2018 at https://www.gazeta.ru/business/2016/06/16/8312309.shtml.

RBC.RU (2018), 'Sechin i Alekperov dali prognozi po tsenam na neft na 2019 god', accessed 25 November 2019 at https://www.rbc.ru/rbcfreenews/ 5c2385139a794712d76abc07.

Rosneft (2011), 'Rosneft dovela uroven utilizatsii poputnogo gaza v OOO RN – Sakhalinmorneftegaz do 95%', accessed 12 October 2018 at https://www.rosneft.ru/ press/releases/item/114239/.

Rosneft (2012a), 'Rosneft i Statoil podpisali soglasheniya po sozdaniyu sovmestnykh predpriyatiy dlya provedeniya geologorazvedochnykh rabot na shelfe', accessed 12 October 2018 at https://www.rosneft.ru/press/releases/item/176919/.

Rosneft (2012b), 'Rosneft i Statoil podpisali Deklaratsiyu o berezhnom osvoyenii rossiyskogo Arkticheskogo shelfa', accessed 12 October 2018 at https://www .rosneft.ru/press/releases/item/18661/.

Rosneft (2012c), 'Na Veninskom uchastke proyekta Sakhalin-3 nachato burenie poiskovo-otsenochnoi skvazhiny', accessed 12 October 2018 at http://www.rosneft .ru/Upstream/Exploration/russia_far_east/sakhalin-3/.

Rosneft (2012d), 'Investitsionnye proyekty Rosnefti po utilizatsii poputnogo neft-yanogo gaza utverzhdeny Ministerstvom economicheskogo razvitia Rossiyskoy Federatsii', accessed 12 October 2018 at https://www.rosneft.ru/press/releases/item/ 177633/.

Rosneft (2013a), 'Rosneft konsolidirovala 100% TNK-BP', accessed 12 October 2018 at https://www.rosneft.ru/press/releases/item/177799/.

Rosneft (2013b), 'Rosneft priobretaet neftegazovye aktivy AK ALROSA', accessed 12 October 2018 at https://www.rosneft.ru/press/releases/item/22666/.

Rosneft (2013c), 'Rosneft i Novatek dogovorilis ob obmene aktivami', accessed 12 October 2018 at https://www.rosneft.ru/press/releases/item/48218/.

Rosneft (2013d), 'Rosneft i Statoil prodvinulis v realizatsii proekta razrabotki trudnoiz-vlekaemykh zapasov nefti v Samarskoi oblasti', accessed 12 October 2018 at https://www.rosneft.ru/press/releases/item/48196/.

Rosneft (2013e), 'Rosneft i ExxonMobil zavershili formirovaniye SP dlya razrabotki trudnoizvlekayemykh zapasov nefti Zapadnoi Sibiri', accessed 12 October 2018 at https://www.rosneft.ru/press/releases/item/48200/.

Rosneft (2014), 'Rosneft i BP podpisali soglasheniye po razrabotke domanikovykh otlozheniy', accessed 12 October 2018 at https://www.rosneft.ru/press/releases/item/153084/.

Rosneft (2015), 'Rosneft i PDVSA podpisali Memorandum v otnoshenii razvitiya prioritetnykh proyektov', accessed 12 October 2018 at https://www.rosneft.ru/press/releases/item/174167/.

Rosneft (2016a), *Annual Report 2016*, accessed 11 October 2018 at https://www.rosneft.com/upload/site2/document_file/a_report_2016_eng.pdf.

Rosneft (2016b), 'Rosneft priobretayet dolyu v krupneishem gazovom mestorozhdenii Sredizemnogo morya', accessed 12 October 2018 at https://www.rosneft.ru/press/releases/item/185061/.

Rosneft (2016c), 'Rosneft uspeshno zakryla sdelku po prodazhe 11% AO Vankorneft ONGC Videsh Limited', accessed 12 October 2018 at https://www.rosneft.ru/press/releases/item/184363/.

Rosneft (2016d), 'Rosneft uspeshno zakryla sdelku po prodazhe 23.9% AO Vankorneft konsortsiumu indiiskikh kompaniy', accessed 12 October 2018 at https://www.rosneft.ru/press/releases/item/183891/.

Rosneft (2016e), 'Rosneft zavershayet formirovaniye mezhdunarodnogo konsortsiuma na baze OOO Taas-Yuryakh Neftegazodobycha', accessed 12 October 2018 at https://www.rosneft.ru/press/releases/item/183889/.

Rosneft (2016f), *Sustainability Report 2016*, accessed 12 October 2018 at https://www.rosneft.com/upload/site2/document_file/RN_SR_2016_EN(2).pdf.

Rosneft (2017a), *Sustainability Report 2017*, accessed 20 January 2019 at https://www.rosneft.com/upload/site2/document_file/RN_SR2018_eng_web_1.pdf.

Rosneft (2017b), 'Rosneft nachala bureniye samoy severnoy skvazhiny na rossiyskom shelfe', accessed 12 October 2018 at https://www.rosneft.ru/press/releases/item/186075/.

Rosneft (2017c), 'Company policy on sustainable development', accessed 20 January 2019 at https://www.rosneft.com/upload/site2/document_file/development_policy_eng.pdf.

Rosneft (2017d), 'Rosneft i Respublica Sakha (Yakutiya) rasshiryayut sotrudnichestvo v sotsialnoy sfere', accessed 12 October 2018 at https://www.rosneft.ru/press/releases/item/187609/.

Rosneft (2017e), 'Rosneft uspeshno zakryla strategicheskuyu sdelku po priobreteniyu 49% Essar Oil Limited', accessed 23 January at https://www.rosneft.ru/press/releases/item/187525/.

Rosneft (2018a), 'Shelfovye proyekty', accessed 12 October 2018 at https://www.rosneft.ru/business/Upstream/offshore/.

Rosneft (2018b), *Sustainability Report 2018*, accessed 25 November 2019 at https://www.rosneft.com/upload/site2/document_file/Rosneft_CSR18_EN_Book.pdf.

Rosneft (2019a), 'Mezhdunarodnoe sotrudnichestvo na shelfe RF', accessed 3 January 2019 at https://www.rosneft.ru/business/Upstream/icoorp/.

Rosneft (2019b), 'Proizvodstvenniye rezultaty za 12 mesyatsev i 4 kvartal 2018 goda', accessed 8 February 2019 at https://www.rosneft.ru/press/releases/item/193733/.

Samoilova, N. (1998), 'Privatizatsiya Rosnefti', *Kommersant*, accessed 23 January 2019 at https://www.kommersant.ru/doc/194943.

Shvarts, E., A. Pakhalov, A. Knizhnikov and L. Ametistova (2018), 'Environmental rating of oil and gas companies in Russia: How assessment affects environmental transparency and performance', *Business Strategy and the Environment*, **27** (7), 1023–38.

Solodovnikova, A. (2013), 'Rosneft obnalichila TNK-BP', accessed 12 October 2018 at https://www.kommersant.ru/doc/2155473.

Starinskaya, G. (2015), 'Ne vremya dlya shelfa', accessed 12 October 2018 at https://www.vedomosti.ru/business/articles/2015/06/15/596298-rosneft-otkladivaet -burenie-na-semi-uchastkah-arkticheskogo-shelfa.

Stulov, M. (2016a), 'Bashneft "do pokupki" Rosneftiyu', accessed 12 October 2018 at https://www.vedomosti.ru/business/galleries/2016/10/12/660481-bashneft-do -pokupki-rosneftyu.

Stulov, M. (2016b), 'Igor Sechin poobeshchal za Bashneft bolshe vsekh', accessed 12 October 2018 at https://www.vedomosti.ru/business/articles/2016/10/03/659350 -sechin-poobeschal.

Stulov, M. (2016c), 'Pokupatelyam Rosnefti mogli pomoch Gasprombank i VBPP', accessed 12 October 2018 at https://www.vedomosti.ru/business/articles/2016/12/ 15/669856-pokupatelyam-rosnefti.

Surzhenko, V. and E. Mazneva (2007), 'Chto vyroslo, to vyroslo', accessed 11 October 2018 at https://www.vedomosti.ru/newspaper/articles/2007/05/04/chto-vyroslo-to -vyroslo.

Usov, I., A. Terentyeva, A. Fadeeva and V. Petleva (2015), 'Soyedinennye Shtaty rasshirili sanktsionnye spiski protiv Rossii', accessed 12 October 2018 at https:// www.vedomosti.ru/business/articles/2015/07/30/602923-soedinennie-shtati -rasshirili-sanktsionnie-spiski-protiv-rossii.

Vestifinance (2018), 'Rossiya ostalas krupneishim postavschikom nefti v Kitai v 2017 godu', accessed 23 January 2019 at https://www.vestifinance.ru/articles/96804.

Vinogradova, O. (2005) 'Echo YUKOSa', *Neftegazovaya vertikal*, **6**, 29.

Yeremenko, Y. (2017), 'Sechin protiv Tesla: O pobede uglevodorodov nad electro-mobilyami i tsenakh na neft', accessed 12 October 2018 at http://www.forbes.ru/ biznes/345515-sechin-protiv-tesla-o-pobede-uglevodorodov-nad-elektromobilyami -i-cenah-na-neft.

3. LUKOIL: patriotic cosmopolite

Vagit Alekperov, the President of LUKOIL, was one of only two heads of the Soviet-era oil companies who managed to retain control of their companies – the other being Vladimir Bogdanov, the Director of Surgutneftegas (see the chapter on Surgutneftegas in this book). Yet Alekperov's strategy has been to move with the times and, in many cases, take the lead on forging change in the Russian petroleum sector, unlike Bogdanov, who has sought to resist change as much as possible.

LUKOIL has enjoyed close ties with the Russian authorities and has always sought to maintain a good relationship with the tax collector. The company has been at the forefront of reform in the industry, driving privatization in the 1990s and promoting technological innovation in the 2000s and 2010s, including shale oil extraction. LUKOIL now lobbies for fair opportunities for private firms in offshore development. The company has sought international partnerships and opportunities to establish its upstream and downstream operations abroad while also developing a strong position in the domestic industry. It has responded flexibly to the challenges posed by oil price volatility and sanctions.

LUKOIL seeks to maintain its reputation as a good corporate citizen that engages in various social investment activities in the areas where it operates. In Russia, this is largely via agreements negotiated with local authorities. The company has committed itself to several domestic and international initiatives that require high levels of transparency and reporting to international standards. LUKOIL measures and monitors its climate change impact, promoting APG utilization in its operations, developing new climate change mitigation technologies and supporting renewable energy projects and research.

LUKOIL is Russia's second-largest oil company after Rosneft. It is one of the most internationalized Russian oil companies, operating in over 30 countries, both upstream and downstream. Headquartered in Moscow, LUKOIL is fully privatized, and its most significant shareholders are Vagit Alekperov (25%), Leonid Fedun (10%), IFD Kapital (8%) and LUKOIL Investment Cyprus (9.2%) (Kommersant 2017a).

Having produced 87 million tonnes of oil in 2018 in total, LUKOIL currently accounts for 12% of proved reserves, 16% of crude production and 15% of refining in Russia. Globally, the company accounts for 1% of proved oil reserves, 2% of production and 2% of refining (LUKOIL 2018a). LUKOIL has four refineries, two mini-refineries and four gas-processing plants in Russia, as well as four refining and petrochemical facilities outside the country (LUKOIL 2018b). It also owns 5390 petrol stations in Russia and 18 other countries (LUKOIL 2018c).

CORPORATE HISTORY: LUKOIL REDEFINES THE RUSSIAN PETROLEUM INDUSTRY

The 1990s

The state oil enterprise LangepasUraiKogalymneft was established on 25 November 1991 by a decree of the USSR Council of Ministers initiated by the First Deputy Minister of the Oil and Gas Industry, Vagit Alekperov. Alekperov had studied the activities of international oil companies (IOCs), and proposed establishing vertically integrated oil companies in Russia, based on the Soviet oil production associations and using Western majors as a role model. He wrote in his book *Russian Oil* that 'in our opinion, we ought to apply the experience of the Western oil companies that unite enterprises throughout the technological chain from well-head to fuel station' (Alekperov 2011, p. 343).

Initially, LangepasUraiKogalymneft included three Siberian production associations (Langepasneftegaz, Uraineftegaz and Kogalymneftegaz) as well as refineries in Perm, Volgograd, Ufa and Mazeikiu. The open joint stock company (OJSC) LUKOIL was created on 5 April 1993 by Government Decree 299 'On Establishment of the OJSC LUKOIL', incorporating the three production associations and just two of the refineries, Perm and Volgograd, along with petroleum product distributors in several Russian regions. Its privatization programme was approved in the same year, and the first issue of shares was registered (LUKOIL 2018j). Under Government Decree 861 of 1 September 1995 'On Improving the Structure of the OJSC LUKOIL', controlling stakes in nine oil production and marketing and service enterprises in the Urals, Volga and Western Siberian regions were added to LUKOIL's authorized capital (LUKOIL 2018k).

During the 1990s, LUKOIL was the undisputed flagship of the Russian oil industry, thanks to the leadership of Vagit Alekperov. The Russian government and Prime Minister Victor Chernomyrdin promoted key LUKOIL initiatives. Yuri Shafranik, the Minister of Fuel and Energy, said at the annual general meeting of LUKOIL in April 1995 that 'LUKOIL projects have always been supported, are currently supported and will be supported in the future by the authorities' (Kommersant 1996).

During the 1990s, LUKOIL drove change in Russia's oil sector. It was the first Russian oil company to develop a privatization plan, and in late 1993, some 12% of LUKOIL shares were sold at voucher and money auctions. Roughly 8% of the shares were distributed among its executives and personnel, including Alekperov and LUKOIL's Vice President, Leonid Fedun. The government held 80% for three years. The next phase of privatization began in October 1995 when bonds were issued to be swapped for ordinary shares. Once the bonds had been redeemed, the government stake was reduced by 11%.

In December 1995, a 5% state stake was sold at a loans-for-shares auction, and 16.07% of the shares were offered at an investment tender. LUKOIL bought its own shares at the loans-for-shares auction (Hudson 1996). In this manner, LUKOIL completed consolidation in 1995, ahead of other oil companies. LUKOIL was the first among its counterparts to recruit an international auditor, KPMG, in 1994 (Neftegaz.ru 2000) and, in 1995, became the first Russian oil company to have an international major among its shareholders: Atlantic Richfield Company (ARCO) (see the section on partnerships with foreign companies below).

During the 1990s, LUKOIL acted as Russia's petroleum ambassador. It was the first Russian oil company to enter the other former Soviet republics, starting with Kazakhstan and Azerbaijan, and Boris Yeltsin's government appointed LUKOIL the coordinator of all Caspian petroleum projects. LUKOIL was also the first to expand beyond the countries of the former Soviet Union and has been active in the Middle East since the mid-1990s, including in Egypt and Iraq.

Throughout its history, LUKOIL has followed a basic rule: to spend what it earned on new acquisitions in the oil sector (Khrennikov 2007). Thus, in 1997, LUKOIL gained control over the oil production company Arkhangelskgeoldobycha and, in 1999, became the key player in Timan-Pechora through the acquisition of the holding company KomiTEK (including the oil production company Komineft, an oil refinery and a fuel marketing unit). Having bought KomiTEK, LUKOIL pledged to improve the environmental situation in the Komi Republic, which had deteriorated due to persistent leakage of oil from Komineft pipelines, including a major spill in 1994 (LUKOIL 2016b, p. 59). LUKOIL established the Timan-Pechora petroleum province where it worked with ConocoPhillips after they signed a memorandum of understanding (MoU) for the joint development of the region with the American major in 1998.

Prior to the 2000s, LUKOIL operated as a state within the state: the company owned tankers and railroad cisterns, a drilling company, LUKOIL-Bureniye, as well as a bank and an insurance company. In 1997, LUKOIL was the first Russian company to begin the large-scale construction of fuel stations operating under its own brand name in Russia; it also started purchasing fuel stations abroad (see the section on internationalization below). It also entered the petrochemical sector and bought Russian Stavrolen in 1998 (see Poussenkova 2010). In addition, it became involved in non-core business activities. Thus, in

1993, the government adopted the Programme of Revival of the Trade Fleet. As an integral element of this programme, LUKOIL launched the construction of new ice-class tankers – a significant contribution to the restoration of manufacturing in Russia. Between 1997 and 2002, it built five Arctic-class tankers in Germany and Saint Petersburg and five river-sea class tankers in Volgograd.

The 2000s

In the new millennium, LUKOIL continued to pursue the same strategy, although it relinquished its status as Russian oil sector flagship to Rosneft. Vagit Alekperov was quick to discern the implications of the changes in Vladimir Putin's Russia and adjusted his modus operandi accordingly. As Isabel Gorst of the *Financial Times* noted in 2007, 'until recently, LUKOIL's Western critics often complained that the company was a dinosaur, rooted in the operational and management practices of the past. But LUKOIL is, in fact, highly adaptive, surviving over 15 years of tectonic changes in the Russian oil industry' (Gorst 2007, p. 1).

One of LUKOIL's basic rules was not to quarrel with the tax authorities. In 2002, LUKOIL willingly paid USD 103 million that the revenue service was demanding for the use of the so-called Baikonur scheme of trading petroleum products, which made it possible to evade excises and profit tax. In 2005, LUKOIL settled all tax claims up to 2003. Unlike Mikhail Khodorkovskiy, Vagit Alekperov did not have any political ambitions, which helped him avoid serious confrontations with the authorities.

LUKOIL responded to the new market demands of the time. If LUKOIL had previously been the state within the state, in 2002, the company began to divest itself of its non-core assets. This decision was partly because LUKOIL was about to lose its leadership status in oil production to YUKOS and partly because it needed to cut costs; it also needed to satisfy foreign investors who did not appreciate the concept of non-core businesses in the structure of an oil company.

During the 2000s, LUKOIL had uneasy relationships with its foreign partners, entering into and breaking up several partnerships, including with ARCO, BP (which acquired ARCO in 2000) and ConocoPhillips (see the section on partnerships with foreign companies below).

During the 2000s, LUKOIL strengthened its position in the refining sector by buying NORSI-Oil's near-bankrupt Nizhny Novgorod refinery for USD 26 million. It also developed its export capacities by establishing the Varandei oil terminal on the shore of the Barents Sea in 2008 (see the section on Arctic and offshore projects below).

In the early 2000s, LUKOIL became the first Russian company to gain a foothold in Saudi Arabia through a partnership with Saudi Aramco. It strengthened its relationships with the Commonwealth of Independent States (CIS), entering Uzbekistan with a focus on gas production. It also expanded into the foreign downstream, purchasing refineries and fuel stations in Europe, although anti-Russian sentiment made it difficult to achieve a strong position in some countries. Also, LUKOIL became the first and only Russian oil company to establish itself in the United States with a network of fuel stations (see the section on internationalization below).

In 2005, LUKOIL strengthened its position in Kazakhstan by acquiring oil producer Nelson Resources. However, its activities in the early 2000s in Azerbaijan declined somewhat compared to the 1990s. Beyond the CIS, LUKOIL has been active in upstream operations in Latin America since the early 2000s, notably in Colombia and Venezuela. It was one of the first Russian oil companies to begin operations in Africa, including Cote d'Ivoire, Cameroon and Nigeria in the 2000s, mainly working offshore (see the section on internationalization below).

Post-2010

In 2011, LUKOIL's oil production declined for the first time, by 5%, and the company called the year 'the bottom', hoping that things would not get any worse. It also had to write off the costs of dry wells amounting to USD 417 million in 2011, almost twice as much as the previous year (Kommersant 2012a). The misfortunes of 2011 sent a strong alert to LUKOIL's leadership that they needed to expand the company's reserve base and gain access to new assets.

In December 2011, LUKOIL deepened its activities in Timan-Pechora through a partnership with Bashneft in the Bashneft-Polyus joint venture (Kommersant 2012b). However, in May 2012, at the request of a minority shareholder of Bashneft, the Russian extractive industry regulator Rosnedra revoked the licence for Trebs and Titov fields from Bashneft-Polyus, returning it to Bashneft (Vedomosti 2012a). It took the partners almost two years to reverse this decision. Oil production at Trebs and Titov began in 2013 and could reach 4.5 million tonnes per annum at its peak. In 2012, LUKOIL won the competition with Rosneft for the Imilorskoye group of fields in Khanty-Mansi Autonomous District. The company hoped that this victory would help it maintain its levels of oil production in Western Siberia, which had declined by 17.5% over the previous five years (Kommersant 2012c).

LUKOIL currently faces serious exploration and production (E&P) challenges, given the depletion of its mature fields in Western Siberia – its main region of operations. LUKOIL's oil production in Russia began to decline steadily in 2015, and in Western Siberia it fell in 2009, 2015 and 2016, that is to say, when oil prices were low. This poses an additional problem since maintaining oil production volumes in Western Siberia requires the use of costly state-of-the-art technologies.

LUKOIL's prospects in foreign refining improved post-2010 with the purchase of refineries in Italy and the Netherlands in the late 2000s. However, the tax and anti-monopoly authorities in Bulgaria and Romania have brought LUKOIL's refining ventures under close scrutiny. While building up a network of fuel stations in various European countries, anti-Russian sentiments led LUKOIL to divest itself of all its fuel stations in the Baltic States (see the section on internationalization below).

In the 2010s, LUKOIL was winding down its operations in Venezuela and Colombia while entering Mexico at the same time. It continued its engagement in Africa with new activities in Ghana from 2014 onwards; it has also been active in Norway, Romania and Vietnam. In 2016, LUKOIL's gas production increased in both Kazakhstan and Uzbekistan, although it came into conflict with the Kazakh authorities over the Karachaganak project (see the section on internationalization below).

COMPANY PROFILE

Production Strategy

LUKOIL's strategic priority is implementing new projects to increase production. This entails both developing greenfield sites and intensifying production at brownfield sites by using advanced technologies, increasing development drilling and enhancing oil recovery (LUKOIL 2018d). LUKOIL's strategy identifies 'growth projects' and 'traditional projects'. The company believes that its top priority should be to ensure increased oil production from growth projects and to stabilize its volumes in its traditional regions of operation.

Growth projects include the Imilorsko-Istochniy licensing area in Khanty-Mansi Autonomous District, Western Siberia,[1] the Bolshekhetsk depression in Yamal-Nenets Autonomous District in the north of Western Siberia, the Yareg and Usinsk fields in Timan-Pechora as well as the offshore fields in the North Caspian (Filanovskogo, Korchagina and Rakushechnoye) and two Uzbek projects (Kandym-Hauzak-Shady and South-West Gissar). Traditional areas of LUKOIL's activities (Western Siberia, Timan-Pechora, Volga-Urals and Kaliningrad regions) are characterized by well-developed production and social infrastructure with some growth projects.

Appreciating the limitations of LUKOIL's reserve base and production, Vagit Alekperov formulated its E&P goals as follows: 'We want to find a major exploration project in Russia to create a petroleum province as we did with the Caspian Sea in the past. We are considering territories in Krasnoyarsk Krai, Sakha and the Irkutsk region' (Vedomosti 2013a). In line with this goal, in 2016, LUKOIL began 2D seismic testing on the Vostochno-Taimyrksiy plot in Eastern Siberia.

LUKOIL is one of the few vertically integrated oil companies in Russia that has a relatively well-balanced ratio between oil production and refining, and it has always made an effort to develop its refining segment. In 2016, LUKOIL commissioned a complex for the deep refining of vacuum gas-oil (the largest in Russia) at the Volgograd refinery. This complex became the final key facility in the large-scale upgrading that LUKOIL launched in 2008. Thanks to this programme, LUKOIL became the first oil company in Russia to make a full transition to production of Euro-5 motor fuels.

LUKOIL has always emphasized the development of its gas business. In 2018, the company's total commercial gas output was 36 billion cubic metres, indicating significant growth from 20 billion cubic metres in 2016. In Russia, LUKOIL's gas production is based on the fields of Bolshekhetsk depression. Its key producing field there is Nakhodkinskoye, which accounts for some 60% of LUKOIL's gas output in Russia. Gas is delivered to the Yamburg gas compressor station and is later pumped through Gazprom's pipelines.

In 2018, international projects made a significant contribution to LUKOIL's total gas production: its output increased to 16 billion cubic metres accounting for 47% of its gas output (see the section on internationalization below).

LUKOIL processes gas at its processing plants in Western Siberia, Timan-Pechora and the Volga regions, and also at its Perm refinery and the Stavrolen petrochemical facility. In 2016, volume of gas processed rose by 6.6% to 4 billion cubic metres (LUKOIL 2018e). In 2016, LUKOIL commissioned a 2.2-billion-cubic-metre facility at Stavrolen that would enable the processing of North Caspian gas into liquids and commercial gas (LUKOIL 2016a, p. 23).

In the gas business, LUKOIL demonstrates its typical pragmatism by cooperating with Gazprom. The two companies signed a strategic agreement in 2005 under which LUKOIL committed to selling its gas to Gazprom at the well-head. The agreement also envisaged the joint participation of the two companies in the exploration and development of new fields in Russia and Uzbekistan. In 2014, Gazprom and LUKOIL extended their strategic agreement for ten years: LUKOIL would continue selling to Gazprom all the natural gas it produced in Russia, regardless of whether volumes and prices changed. LUKOIL does not want to trade its gas in Russia independently mainly because of difficulties with access to the Unified Gas Supply System.

Unlike Rosneft and Novatek, which have been trying to lure customers away from Gazprom, LUKOIL does not have the lobbying power that they have and prefers to treat Gazprom as an ally rather than as a competitor. However, in February 2017, LUKOIL asked the government for the first time to include 'liberalization of the gas market as well as access to the continental shelf' among the state energy policy priorities aimed at increasing competitiveness within the sector.

LUKOIL differs from its counterparts in its focus on petrochemicals. It owns petrochemical facilities in Russia (Saratovorgsyntez, Stavrolen), Bulgaria (Neftokhim Burgas) and Italy (ISAB). LUKOIL meets domestic Russian demand for certain chemical products and exports chemicals to some 30 countries (LUKOIL 2018f). The company has also successfully engaged itself in power generation, particularly in southern European Russia (providing 99% of power consumed in the Astrakhan region and 62% in Krasnodar Krai). LUKOIL's power generation business is represented by a whole vertically integrated chain from generation to transmission and sales of heat and electricity to external consumers. The total capacity of LUKOIL's power-generating facilities amounts to 5.8 GW and is made up of 73% commercial generation and 27% supporting generation.

LUKOIL's Role in the Arctic and Offshore Oil Extraction

Since 2008, Russian law has restricted offshore exploration and production licences to experienced state-owned companies (Rosneft and Gazprom). The restriction applies to areas more than 12 nautical miles from the coast, that is to say, beyond territorial waters. LUKOIL was already operating in Russia (the Caspian and Baltic Seas) prior to 2008. It was allowed to keep these projects but has not been able to apply for any new licences (The Barrel 2015).

Vagit Alekperov has long argued for the equal rights of private and state oil companies to carry out offshore projects in Russia. He has complained that the state selects only those 'who have the right' rather than the experience and technologies (Vedomosti 2012b). Arguing for the liberalization of access to the Russian continental shelf, Alekperov said that 'LUKOIL is a Russian national company; it is a taxpayer in Russia; the state authorities control it on equal terms with the state companies. We follow the same laws; we are inspected by the same authorities' (Vedomosti 2013a).

The Russian authorities are now considering opening up offshore reserves to private companies mainly because state-owned companies have found it difficult to meet their licence obligations in an era of low oil prices and Western sanctions. While LUKOIL would welcome reform, the same challenges could also undermine their own ambitions (The Barrel 2015).

Despite the regulatory limitations and practical challenges, LUKOIL is one of the few Russian oil companies that has accumulated a significant amount of experience of offshore operations in both Russia, where it tends to work alone, and abroad, where it is usually part of international consortia (see the section on internationalization below for more on its international projects).

LUKOIL has been leading operations on the Russian continental shelf for over 20 years. It began large-scale exploration in the Caspian Sea in 1995. Since then, it has discovered ten new fields in the Russian sector of the Caspian Sea, with total recoverable reserves of 1.1 billion tonnes of oil equivalent.

In April 2010, LUKOIL commissioned its first project in the Russian sector of the Caspian Sea – the Yuri Korchagin field (discovered in 2000). This field yielded 1.6 million tonnes of oil in 2015. The Filanovskogo field (discovered in 2005) was commissioned in 2016. This was the largest oil field in the Russian sector of the Caspian Sea, with 80 times the average output of LUKOIL wells, and with recoverable reserves of oil totalling 129 million tonnes and 30 billion cubic metres of gas. Taking into account the environmental vulnerability of the Caspian Sea, LUKOIL aims for zero discharges from the Filanovskogo field and deposes all waste onshore (LUKOIL 2016a, p. 4). LUKOIL's other Caspian fields include Kuvykin and Rakushechnoye, which are due to be commissioned in 2022 and 2026, respectively.

Overall, the Caspian Sea has brought LUKOIL many disappointments in terms of the discovery of commercially viable hydrocarbon reserves, such as the Karabakh project in the Azeri sector, the Tyub-Karagan field in the Kazakh sector (Vedomosti 2008a) and the Yalama field in the Azeri sector (Vedomosti 2009a). Nonetheless, the Caspian Sea remains LUKOIL's most profitable region. Its projects in the Russian sector of the Caspian offer the greatest fiscal benefits (low rates of mineral production tax and zero export duty). Therefore, the Caspian Sea is set to become one of the key sources of LUKOIL's income (Vedomosti 2016a). LUKOIL has hoped to stem the decline of its Russian production through its Caspian projects. Its leadership always reminds the government about its achievements in the Caspian when it tries to convince the authorities to remove constraints on private companies for offshore operations.

LUKOIL is also active in the Baltic Sea. In 1997, it started developing the Kravtsovskoye field, which produced its first commercial oil in 2004. This was the first purely Russian offshore project that entered the commercial phase. In 2015, LUKOIL-Kaliningradmorneft made several small discoveries in the Baltic Sea. In 2016, LUKOIL tried to obtain another plot – Nadezhda – in the Baltic Sea. This is located within the 12 nautical mile zone, that is, in territorial waters; therefore, private oil companies can hope to secure access to the reserves.

LUKOIL has yet to develop offshore operations in Arctic waters. However, much of LUKOIL's production activity in the Timan-Pechora province is located north of the Arctic Circle in Nenets Autonomous District and the northern Komi Republic. Therefore, the company is familiar with working in onshore Arctic locations and has made efforts to develop the related infrastructure. With rising oil production and a lack of essential infrastructure in northern European Russia, LUKOIL built the Varandei oil terminal on the shore of the

Barents Sea – the world's first 12-million-tonne-per-annum ice-resistant oil pier operating all year round in the Arctic. It is intended for marine transportation of exported oil that LUKOIL produces in Nenets Autonomous District. Launched in 2008, the terminal works on a 'zero-discharge' basis (Alekperov 2011, p. 384).

Partnerships with Foreign Companies

LUKOIL's first foreign partnership was established in the mid-1990s: in 1995 and 1996, the American company ARCO bought securities worth USD 250 million and became a shareholder of LUKOIL, with 7.99% of the shares. In 1997, the two companies formed the joint venture LukArco to seek opportunities within and outside Russia. The joint venture was unable to obtain any assets in Russia but acquired 5% of TengizChevroil in Kazakhstan and 12.5% of the Caspian Pipeline Consortium.

During the 2000s, LUKOIL had uneasy relations with its foreign partners. Following the acquisition of ARCO in 2000, BP became the owner of 7.99% of LUKOIL. However, in January 2003, BP sold this stake (LUKOIL 2018g), and in 2009 BP withdrew from the joint venture LukArco because of a disagreement with Caspian Pipeline Consortium partners. LUKOIL bought out its 46% share for USD 1.6 billion and became the sole proprietor of LukArco (Vedomosti 2009b).

In September 2004, ConocoPhillips bought the 7.59% stake in LUKOIL (the last stake in the company owned by the state) for USD 1.9 billion and, afterwards, gradually increased its share to 20.86%. The Kremlin was satisfied with the deal, and Alekperov was awarded the Order for Service to the Fatherland of the fourth degree in 2005 (Neftrossii.ru 2014). ConocoPhillips was supposed to help promote LUKOIL's projects in the West. Training of LUKOIL's personnel in ConocoPhillips's offices in Houston was another important aspect of the deal, while 15 leading experts of ConocoPhillips worked as advisors at LUKOIL's headquarters.

In 2005, ConocoPhillips (30%) and LUKOIL (70%) established Naryanmarneftegaz to develop the northern part of the Timan-Pechora basin, primarily the large Yuzhno-Khylchuyusk field launched in 2008. It was hoped that this asset would help combat LUKOIL's oil production decline in Western Siberia (Vedomosti 2008b). However, these ambitions were not realized because of a serious error made by the geologists. In 2008, the proved reserves of Yuzhno-Khylchuyusk had been estimated at 70 million tonnes; however, oil production there fell from 7 million tonnes in 2009 to 3.3 million tonnes in 2011, and its reserves were downgraded to 20 million tonnes in 2011 (Kommersant 2012a).

In March 2010, ConocoPhillips announced that it wanted to sell its stake in LUKOIL. In August, LUKOIL's subsidiary bought 7.59% for USD 3.44 billion. Leonid Fedun explained the purchase thus: 'We considered it our patriotic duty that these shares should return to Russia' (Vedomosti 2010a). Seemingly the company received a signal from the Kremlin that outside investors should not be allowed to acquire the stake. After that, LUKOIL purchased the remaining shares (Vedomosti 2010b). In 2012, LUKOIL also bought out ConocoPhillips's shares in Naryanmarneftegaz, thus ending its collaboration with ConocoPhillips. Alekperov stated that after ConocoPhillips sold its stake in LUKOIL, LUKOIL's chances of receiving access to reserves would improve because it no longer had a strategic (foreign) partner with 20% (Neftegaz.ru 2010).

Internationalization

LUKOIL was the first Russian oil company to internationalize aggressively. It started expanding abroad during the 1990s and continued the trend in the 2000s and 2010s. Flexibility and rationality characterize LUKOIL's globalization policy. Thus, the company has not internationalized just for the sake of internationalization, and it has been disciplined about selling assets that have proved to be commercially unsatisfactory. However, the company's Russian roots have sometimes diminished its options, particularly when it has attempted to enter the European downstream sector, where it continued to be regarded with suspicion as a possible tool of the Kremlin despite being a wholly privately owned company.

The reasons for LUKOIL's globalization were multifaceted: a need to increase its reserve base and supplement declining production in Western Siberia, a desire to get access to offshore projects denied to it in Russia after 2008 and the intention to reduce the fiscal burden. LUKOIL's foreign ventures also allowed the company to learn from its international partners' modern management techniques. Alekperov noted: 'We have consortia with Exxon, Shell, BP and Chevron in many regions of the world. We have adapted to this system of operations when everybody bears his share of responsibility' (Neft Rossii 2016).

In the 1990s, LUKOIL began work at the Tengiz and Kumkol fields in Kazakhstan. In 1993, it signed its first framework agreement with SOCAR – the state oil company of Azerbaijan. In 1994, the so-called contract of the century was concluded for Azerbaijan's Azeri-Chirag-Guneshli fields in which LUKOIL held 10% (Newsru.com 2002). LUKOIL subsequently expanded to other fields in Azerbaijan: Shakh-Deniz, Yalama and Karabakh. At the time, it was vitally important for Russia to re-establish its influence in the post-Soviet space and restart energy dialogues with CIS countries, and LUKOIL played a key role in steering Russia's foreign policy towards Azerbaijan and Kazakhstan (Gorst and Poussenkova 1998). In the early 2000s, LUKOIL continued its strategy of strengthening its position in the former Soviet republics, entering the upstream of Uzbekistan, which became an important part of LUKOIL's gas business.

The Middle East is another strategic region for LUKOIL, with operations in Saudi Arabia, Iraq and Egypt. The Meleiya project in Egypt, which LUKOIL joined in 1995, was one of its first overseas E&P ventures and continues to be productive. Currently, LUKOIL is involved in this project and in two other projects in Egypt. Iraq was also a focus of LUKOIL's globalization efforts from the start. In 1997, LUKOIL bought a controlling interest in the Russian consortium established to operate the West Kurna-2 deposit in Iraq. It also formed the joint venture LukAgip with the Italian Agip to develop petroleum fields in Tunisia, Egypt and Libya (in 2005, LUKOIL bought Agip out of the joint venture). In a significant development, LUKOIL was the first Russian company to gain a foothold in Saudi Arabia with the LUKSAR joint venture, which was established in 2004 between LUKOIL (80%) and Saudi Aramco (20%).

What was unusual at the time was that LUKOIL entered the foreign downstream sector. Taking advantage of the global economic downturn after the 1998 financial crisis, LUKOIL bought refineries in Bulgaria, Romania and Ukraine at knock-down prices. LUKOIL Neftokhim Burgas, the only refinery in Bulgaria, accounts for some 70% of wholesale deliveries of fuel in the country. LUKOIL, with its over 200 petrol stations in Bulgaria, controls roughly 26% of the market.

Table 3.1   *LUKOIL's fuel retail acquisitions in Europe*

Year	Country	LUKOIL acquisitions	Number of stations
2003	Serbia	79.5% share in Serbia's Beopetrol for EUR 117 million	200
2005	Finland	Oy Teboil Ab and Suomen Petrooli Oy for USD 160 million	289
2006	Hungary	15 petrol stations in Hungary from Austria's ABA, bringing the number of its fuel stations in Hungary to 26 units	26
2006	South-Eastern Europe	Joint venture with Slovenian Petrol (51%) to deal with retail sales of petroleum products in South-Eastern Europe (Vedomosti 2006a)	
2007	Six European countries	Petrol stations in six European countries from ConocoPhillips for USD 442 million	376
2012	Belgium/ Netherlands	LUKOIL's Zeeland refinery (formerly TRN) supplies fuel to its petrol stations in Belgium and the Netherlands (Vedomosti 2012c).	13 + 46

Source: LUKOIL website.

LUKOIL was the first and only Russian oil company to gain a foothold in the United States. It did so with the acquisition of Getty Petroleum Marketing, with 1300 fuel stations in 2000, and the purchase of a network of 795 Mobil-branded fuel stations in New Jersey and Pennsylvania from ConocoPhillips. By mid-2005, LUKOIL controlled 8% of the retail fuel market in these regions (Vedomosti 2005a). However, in December 2007, LUKOIL began ridding itself of the Mobil fuel stations (Energyland 2008), and in early 2011 it sold Getty Petroleum. As a result, its number of US fuel stations fell from 1334 to 635 in 2011 (Vedomosti 2011) and 285 in 2016.

During the 2000s, LUKOIL failed to obtain any refining facilities in Europe, which forced it to focus on developing its Russian refining assets (Neftegaz. ru 2007). In 2002–03, the Greek government blocked LUKOIL's attempt to negotiate a 23% stake in the Greek state company Hellenic Petroleum with its three oil refineries and 1500 fuel stations. The trade unions had strongly opposed the proposal, and the government deemed it 'unacceptable from the point of view of national interests' (Gaiduk and Lukin 2006). In 2002, LUKOIL's attempts to purchase Polish refineries and gas stations failed for political reasons (Neft i kapital 2002). LUKOIL was also unable to acquire Mazeikiu Nafta in Lithuania. In 2006, LUKOIL reached an agreement with Kuwait Petroleum International to buy a refinery in Rotterdam; however, its owners changed their mind at the last moment (Energetika i Promyshlennost Rossii 2008). In 2007, LUKOIL wanted to acquire a share in the Czech refinery Ceska Rafinerska from ConocoPhillips; however, the Czech government was unwilling to see a Russian company among the shareholders and preferred to work with ENI. In 2008, LUKOIL's attempt to purchase a 30% stake in Spanish Repsol was approved by the Spanish Prime Minister; however, the opposition parties and the media launched a campaign warning that LUKOIL would act as an 'arm of the Kremlin' and hinting about connections with the Russian mafia. Ultimately, LUKOIL had to abandon its plans (Forbes 2009).

In mid-2008, LUKOIL finally made a breakthrough into west European refining. It established a joint venture with the Italian company ERG, to which ERG contributed its 16-million-tonne-per-annum ISAB refining complex in Sicily. LUKOIL purchased 49% of the shares of the two refineries in the complex, allowing it to double its European refining capacity (Vedomosti 2008c). The deal was closed in December 2008 despite the financial crisis. In 2009, LUKOIL bought 45% of the Total Refinery Netherlands (TRN) from Total (Vedomosti 2009c). These two acquisitions helped LUKOIL improve the balance between its production and refining volumes. In 2013, LUKOIL bought another 20% of ISAB for some EUR 400 million; it planned to refine Iraqi oil, primarily from West Kurna-2 (Vedomosti 2013b).

LUKOIL also had trouble with the refineries it had bought in the 1990s, mainly because of the general anti-Russian sentiment in Eastern Europe.

In 2011, in Bulgaria, LUKOIL was suspected of fraudulent export because LUKOIL's Neftochim Burgas refinery did not install fuel metering devices on time. As a result, officials closed storage facilities in Burgas and revoked the licence of Neftochim Burgas (Kommersant 2011a). LUKOIL managed to settle the problem in court, but almost immediately, the Bulgarian authorities began to accuse the company of violating anti-monopoly legislation and launched investigations of LUKOIL and its trading division. In 2014, the Romanian authorities investigated LUKOIL's Petrotel refinery accusing the company of tax evasion and money laundering, although the charges were dropped in 2016 (Kommersant 2017b).

Table 3.2   LUKOIL's network of petrol stations (number of stations at year-end)

	2012	2013	2014	2015	2016
Russia	2368	2424	2481	2544	2603
Western Europe	2420	2427	2322	2336	2177
Baltic States	212	210	211	142	0
CIS	503	479	469	245	244
United States	425	327	299	289	285

Source: LUKOIL (2018j).

LUKOIL's foray into the European fuel retail market was successful because, unlike refineries, the fuel stations were not considered strategic enterprises. In 2003, the company purchased 200 fuel stations in Serbia, followed by 289 units in Finland in 2005 and afterwards in Hungary, Slovenia and other European countries in the mid-2000s. In 2012, the company made further acquisitions in Belgium and the Netherlands (Table 3.1). At the same time, in 2012, LUKOIL decided to sell 85 fuel stations in Serbia because of their low profitability (Kommersant 2012d). In December 2015, Vagit Alekperov admitted that LUKOIL wanted to withdraw from the Baltic States because of strong anti-Russian sentiments there (Vedomosti 2016b); in 2016, the company divested itself of all its fuel stations in the region (Table 3.2).

   In 2005, LUKOIL also strengthened its position in Kazakhstan by acquiring 65% of the oil-producing company Nelson Resources for USD 2 billion (Vedomosti 2005b). In October 2005, Vagit Alekperov met Nursultan Nazarbayev, who announced that the Kazakh leadership had approved of LUKOIL's activities in Kazakhstan in general and of its acquisition of Nelson Resources in particular (Vedomosti 2005c). However, LUKOIL encountered serious problems with CNPC in Kazakhstan about Turgai Petroleum and its acquisition of Nelson Resources; the two companies were forced to settle their differences by arbitration (Vedomosti 2010c). In 2006, Vagit Alekperov commented that LUKOIL is one of the major players in Kazakhstan, where it produces some 6 million tonnes and participates in six big projects.

> But one has to note that the Chinese have more opportunities than we do. For example, we could not offer PetroKazakhstan the price that CNPC paid for the company. It is difficult for us to compete with them because we work under market conditions while the Chinese national company has access to cheap and stable government money. (Vedomosti 2006b)

Despite LUKOIL's generally amicable relations with the Kazakh government, it had problems with Karachaganak, as did other consortium members. In April 2016, the Kazakh authorities demanded USD 1.6 billion from the consortium for violating the provisions of its production-sharing agreement (PSA). Similar demands made in 2010, when the consortium was accused of tax evasion and oil production above established levels, had resulted in the transfer of a 10% stake in Karachaganak to Kazakhstan.

LUKOIL's projects in Central Asia helped it develop its gas business. In 2016, due to the successful implementation of South-West Gissar and Kandym-Hauzak-Shady projects, LUKOIL's gas production in Uzbekistan rose by 8%. LUKOIL's gas production also increased in Kazakhstan by 8% at the Tengiz field and by 16% at the Karachaganak field.

Compared to the 1990s, the 2000s witnessed the relative weakening of LUKOIL's position in Azerbaijan. In 2003, it sold its stake in the Azeri-Chirag-Guneshli project to the Japanese company INPEX Southwest Caspian Sea for USD 1.3 billion. Alekperov explained that this decision was part of LUKOIL's restructuring programme and announced that the company 'would continue to consider opportunities to withdraw from the projects where it does not have operatorship' (Ria Novosti 2002). Currently, it works mainly on the Shah-Deniz gas condensate field located in the Azeri sector of the Caspian Sea.

LUKOIL has been active in upstream operations in Latin America since the early 2000s. Recently, it decided to wind down its operations in Colombia and Venezuela while entering Mexico at the same time (see Table 3.3).

LUKOIL was one of the first Russian oil companies to begin operations in Africa, entering Cote d'Ivoire in 2006, followed by Cameroon (2008), Nigeria (2009) and Ghana (2014). In these countries, LUKOIL has mainly worked offshore – the area that has been largely denied to it in Russia. Vagit Alekperov said that LUKOIL made a bet on Africa as a promising petroleum region (Kommersant 2012e). The company was particularly active in Africa in 2014, discovering seven hydrocarbon fields in one block in Ghana while also joining projects in Cameroon and Nigeria. LUKOIL believes that deep-water projects in West Africa can be very promising but only with higher oil prices. Therefore, the company has decided to adopt a 'wait-and-see' policy on offshore projects, such as Tano in Ghana, OML-140 in Nigeria and Etinde in Cameroon (Kommersant 2017c).

LUKOIL's only activities in Asia are currently in Vietnam where, in 2011, LUKOIL Overseas bought a 50% stake in an offshore block. As far as European upstream is concerned, LUKOIL is active in Norway where, in 2013, it was one of the winners of the 22nd licensing round. A minor gas field was discovered in 2015 (Kommersant 2013a), but because of its size, by 2016, LUKOIL was contemplating withdrawing from the block as it was dissatisfied with the economics of the project (Kommersant 2016a). LUKOIL entered Romania's upstream in 2011 and managed to discover considerable reserves within the Trident block where it continued exploratory drilling.

Transparency

In 2011, Transparency International and Revenue Watch released a review of the policies of 44 companies responsible for 60% of the world's oil production (Transparency International 2011). From Russia, only LUKOIL, Rosneft and Gazprom were present. The report covered three key areas. The first was 'Reporting on anti-corruption programmes', where LUKOIL scored poorly with 9%, ranking below Rosneft (41%) but above Gazprom (0%). Second, under 'Organisational disclosure', LUKOIL scored a respectable 50%, although ranking below both Rosneft (75%) and Gazprom (81%). The third area was 'country-level disclosure of foreign operations', where most participating countries scored poorly (average 16%), rendering LUKOIL's 15% a near-average score (Rosneft and Gazprom did not feature prominently on the list).

In the 2011 report from Transparency International and Revenue Watch, LUKOIL was also identified as a non-supporter of the Extractive Industries Transparency Initiative (EITI), which was perceived as an indication of a company's transparency credentials. To date, neither LUKOIL nor any other Russian company has been listed as a supporter of EITI. The Russian government has not signed up to EITI; however, some of the countries where LUKOIL has its operations have done so (EITI 2018a, 2018b).

Nonetheless, LUKOIL has over the years displayed a growing tendency towards greater transparency and readiness to participate in international initiatives with reporting and disclosure requirements. In 2002, for instance, LUKOIL was the first Russian company to receive a full listing on the London Stock Exchange, which has substantial disclosure requirements (London Stock Exchange 2018). The company is also listed on the Frankfurt, Munich and

Stuttgart stock exchanges and on the US OTC market. In a 2010 interview with EURACTIV, Alekperov stated that 'LUKOIL is a private company quoted on the London Stock Exchange, and our Western partners see us as any other company working under the principle of transparency in business affairs' (EURACTIV 2018).

LUKOIL produced its first sustainability report in 2005. The sustainability report is now published annually in line with GRI guidelines (GRI 2018) (see the section on CSR below). In 2008, LUKOIL became a member of the UN Global Compact, which requires the participating companies to submit an annual 'Communication on Progress', which is published on the Global Compact website (United Nations Global Compact 2019). Since the start of its reporting in 2009, LUKOIL has progressed from 'Learner' level to 'Active' level in its progress communications (on a par with BP). The highest level is 'Advanced', achieved by Equinor and Shell, among others. Since 2013, LUKOIL has been participating in the Carbon Disclosure Project, which involves disclosing information on its GHG emissions (see the section on climate policy below).

In 2014, WWF Russia and CREON Energy released a rating of the environmental performance of Russian oil and gas companies, including performance on disclosure and transparency. The rating was subsequently updated in 2016, 2017, 2018 and 2019. In 2014, LUKOIL came ninth behind Rosneft (seventh) and Gazprom (third) (Surgutneftegas was ranked the highest). In 2016, LUKOIL came fourth and, by 2017, had worked its way up to second place behind Rosneft but one place ahead of Surgutneftegas, while Gazprom had fallen to sixth place. In the area concerning 'Disclosure and transparency', LUKOIL came in joint eleventh in 2014 and joint third in 2016 but fell to joint sixth in 2017. In 2018 and 2019, LUKOIL achieved first place.

These ratings suggest that LUKOIL has managed to improve its transparency and disclosure. The company appears to have been successful in making steady progress in these areas, although observers have identified room for further improvement.

CSR

LUKOIL strives to project the image of a good corporate citizen. The company established a charity fund in 1993 – one of the first of its kind in contemporary Russia (LUKOIL 2018h). It is a member of the Global Compact in Russia (see United Nations Global Compact 2019) and is a signatory to the Social Charter of Russian Business (Russian Union of Industrialists and Entrepreneurs 2019). LUKOIL has adopted a unique corporate social code, which, among other things, determines its policy on local communities living nearby its operation sites (LUKOIL 2019a). LUKOIL holds the ISO 14001 standard for environmental management systems. It has established a Health, Safety and Environment Committee, accountable directly to LUKOIL's president (LUKOIL 2016b). LUKOIL has also committed itself to support the UN's 2030 agenda for sustainable development (LUKOIL 2016b), and LUKOIL's *Sustainability Report 2017* was produced in line with the 2030 agenda goals (LUKOIL 2017).

LUKOIL's sustainability report is now published annually in line with the GRI guidelines (see Global Reporting 2019). In 2017, LUKOIL won the Moscow Exchange Award for the 'Best Corporate Social Responsibility and Sustainability Report'. The company has repeatedly been ranked among the top ten companies in Russia for 'Responsibility and Openness' by the Russian Union of Industrialists and Entrepreneurs, which produces sustainability indexes based on the publicly available reports of Russia's 100 largest companies.

Alekperov's statements at the beginning of LUKOIL's sustainability reports reflect the evolution of LUKOIL's thinking about CSR. Alekperov's address at the beginning of the 2011–12 report focused primarily on LUKOIL's commercial achievements and technical innovations: 'We consider it our duty to pay our taxes in a timely manner, which helps the state to resolve social challenges in the regions of our Company's presence' (LUKOIL 2012, p. 2). Subsequent addresses exhibit a progressively greater focus on the environment and local communities. The 2016 address emphasized the need to balance economic, environmental and social factors in business development. And in the opening paragraph of his 2017 address, Alekperov stated: 'The company has always sought to work not only to benefit its shareholders and employees but also the society as a whole. We are convinced that our successful development is only possible if we take into account the interests of the communities of the countries where we operate' (LUKOIL 2017, p. 6). This indicates a gradual shift, at least in LUKOIL's public relations department, towards appreciating the need to be more explicit in its public announcements on environmental protection and the well-being of local communities and wider society; the company also acknowledges that its responsibility in this sphere transcends the mere need to pay taxes so that the state can take care of things.

According to the *Sustainability Report 2016*, LUKOIL contributed over RUB 12 billion to charity, sponsorship and social investments (LUKOIL 2016b). The bulk of these funds (RUB 7.3 billion) was channelled via socio-economic partnership agreements with the regions of operations, while RUB 1.8 billion was spent on supporting public funds and organizations. In 2016, LUKOIL signed cooperation agreements with all the Russian regions where it had operations, pledging funds to finance social and sports programmes, construction and maintenance of cultural facilities. In 2018,

LUKOIL provided social support worth around RUB 9 billion in the regions where it operated (LUKOIL 2018l).

LUKOIL's social code sets out how it should interact with indigenous peoples. The provisions are implemented through annually approved programmes in its regions of operation where indigenous peoples live, namely Khanty-Mansi Autonomous District, Yamal-Nenets Autonomous District, Nenets Autonomous District, Komi Republic and Krasnoyarsk Krai. LUKOIL's support for indigenous peoples is provided primarily through agreements between LUKOIL and regional authorities, licensing agreements and agreements on the socio-economic development of the regions where indigenous peoples live. One of the most efficient forms of support is economic contracts on cooperation with heads of the territories inhabited by indigenous peoples. These agreements define the procedure for the use of land, compensation, educational support and medical services. Support provided via such agreements in Khanty-Mansi Autonomous District, for instance, includes cash payments, boats, snowmobiles, boat engines and generators. LUKOIL also signs agreements on cooperation with civil society organizations that represent the interests of indigenous peoples.

LUKOIL coordinates its schedules for field development and exploration activities with indigenous peoples at the planning phase to avoid future conflicts. However, despite the availability of officially formulated policies on local inhabitants, conflicts do occur. For instance, in 2014, a controversy broke out in the Izhma District of the Komi Republic when the citizens unanimously voted for the termination of LUKOIL-Komi's activities in the region, among other things because the company had begun drilling near the village Krasnobor without having notified the local citizens and administration and obtained the requisite permits (Greenpeace 2014). The local population had previously been affected by repeated spills from the ageing oil pipelines in the region, which had undermined LUKOIL-Komi's social licence to operate (Kelman et al. 2016).

LUKOIL tries to regulate the behaviour of the employees of its subsidiaries to avoid conflicts with indigenous peoples. LUKOIL Western Siberia is governed by Order 262 On Measures to Limit Access to Territories of Communal Lands. Departments for interaction with indigenous peoples are established in subsidiaries so that they can convey their complaints and proposals to the company (LUKOIL 2016b, pp. 92–3).

Support for education is another component of LUKOIL's CSR activities. The company has been sponsoring scholarships for talented students specializing in petroleum, petrochemicals and power generation. For instance, in 2016, 185 students across Russia received LUKOIL corporate scholarships. In 2018, 190 students and 79 teachers received scholarships with a total value of RUB 9 million (LUKOIL 2018l). LUKOIL also sponsors several specialized medical research and development centres and contributes to the development of the system of medical services in the regions of its operations. In 2016, it assisted the Russian Cardiological Scientific Centre, Institute of Surgery, Mezensk Central Regional Hospital, Limansk Regional Hospital, Ukhta Children's Hospital, Perm Children's Clinical Hospital and so on (LUKOIL 2016a, p. 77). The company supports the arts, cultural and historical heritage, religious traditions and spiritual culture, cultural activities, museums, churches and monasteries across the country. Since 2002, LUKOIL has been organizing annual competitions to sponsor social and cultural projects. In 2018, 3645 applications were submitted and 773 projects were financed, and financial support was RUB 141 million (LUKOIL 2018l).

In 2006, Vagit Alekperov established the foundation Our Future to provide support for social entrepreneurship. The foundation helps people create new businesses and holds biannual contests for business ventures throughout Russia (Neftegaz.ru 2010). In 2013, Alekperov organized the first international conference, Social Innovations, in which some 400 people participated. The objective was to create an expert forum where people interested in social entrepreneurship could communicate, promote new ideas and identify the problems (Vedomosti 2013c). 'More than a purchase!' is a partner project of LUKOIL and Our Future and involves shops at LUKOIL fuel stations selling goods manufactured by social entrepreneurs. The first outlet opened in 2014 (LUKOIL 2016b, pp. 88–9).

Innovation

LUKOIL is now more active than ever in applying modern methods of enhanced oil recovery, including hydraulic fracturing (fracking), chemical, hydrodynamic and thermal methods. In 2016, 237 horizontal wells were completed with an average daily output of 79 tonnes per day, including 87 wells with multi-zone fracking. The share of the horizontal wells grew from 22% in 2015 to 28.6% in 2016 (LUKOIL 2016a, p. 53).

LUKOIL has sought partnerships with foreign companies to help develop and apply new technologies, an example of which is its partnership with Total to exploit shale oil reserves in Western Siberia. However, the partnership was halted because of the international sanctions against Russia in connection with the conflict in Ukraine (see the section on the shale revolution below). LUKOIL has also become involved in hard-to-recover reserves overseas to gain experience. For instance, in partnership with Saudi Aramco, LUKOIL carried out an appraisal of tight gas production techniques for the Tuchman and Mushaib fields (Reuters 2014).

COPING WITH CHANGE

Shale Revolution

Thanks to the Bazhenov formation in Western Siberia, Russia has the second-largest shale oil reserves in the world (after the United States). LUKOIL has long been trying to develop the Bazhenov formation. In its *Sustainability Report 2011–2012* (p. 12), the company stated that the currently available technologies will only be sufficient to recover 2–4% of the Bazhenov reserves. It has, therefore, invested in the development and testing of advanced technology through its subsidiary RITEK (Russian Innovative Fuel and Energy Company). RITEK is already active in developing advanced methods for extracting oil from the Bazhenov formation, including a system of the thermo-gas impact that is aimed at heating the surrounding rock (Alekperov 2011, pp. 387–8). RITEK applied the technology of thermo-gas impact at test sites for the first time in 2009, hoping that this technology would permit increasing the oil recovery ratio by 10–30% (Oilgas.com 2017). The second test site was commissioned in 2015. In 2017, RITEK drilled one well at a third testing site (Neftegaz.ru 2017).

LUKOIL has sought foreign partners with relevant skills and state-of-the-art technologies to work on the Bazhenov formation. In 2014, LUKOIL (51%) and Total (49%) signed an agreement creating a joint venture to explore and develop the tight oil potential of Bazhenov (Vestifinance.ru 2014). The partners planned to begin seismic testing in 2014 and exploration drilling in 2015. However, the activities were halted because of the anti-Russian sanctions that targeted the development of shale reserves. Total hoped that they could return to the project as soon as the sanctions were lifted. Without a foreign partner, LUKOIL would likely be able to begin only test production at the Bazhenov formation. Alekperov said that LUKOIL appreciated the difficulties that Total faced and that 'we undertook this mission, and we'll finance this programme' and that the French major could return to the project compensating for LUKOIL's costs incurred during its absence (TASS 2015).

Oil Price Volatility

LUKOIL, and Vagit Alekperov in particular, have always monitored global oil prices closely. The company has made efforts to forecast them and has developed its budgets proceeding from realistic and sometimes even conservative scenarios. It is noteworthy that LUKOIL produces an energy market outlook on a regular basis called 'Main Trends in the World Oil Market Development up to 2030' (LUKOIL 2018i).

Vagit Alekperov has realistic views about economic upheavals, as his following opinion about the 1998 crisis shows:

> nevertheless, it was the market system of management that made it possible for the oil sector to survive the crisis. [The year] 1998 was the acid test for the industry when world oil prices dropped to USD 9 per barrel. In this situation, not only did the Russian companies not reduce production, but they did not even freeze their investment programmes. Moreover, the oil price collapse became a driver for the rapid growth of production and financial indicators. It forced the oil companies to drastically improve the efficiency of their activities through cost-cutting, expanding sales volumes and enhancing product quality. (Alekperov 2011, p. 347)

According to Alekperov, the 2008–09 financial crisis allowed LUKOIL to analyse all its mistakes and make unconventional decisions in order to secure the company's long-term stability. This was achieved through cost-cutting and by terminating certain investment projects. Thus, LUKOIL cut its overall costs by 36%, operating costs by 15% and commercial and administrative costs by 15%. It also reduced investments by almost 40% (Vedomosti 2009d). In the autumn of 2008, when Russian companies began downsizing on a large scale, Vagit Alekperov assured his employees that 'he would not allow it'. Yet during the first half of 2009, the number of LUKOIL's personnel declined from 152 500 to 145 600, although staff were only dismissed from the subsidiaries. The headcount in the head office even increased from 2109 people in 2008 to 2196 in 2009, and their salaries grew (Oilcareer.ru 2009).

Vagit Alekperov made the following statement on the 2008–09 crisis:

> The system should be flexible. We were able to weather the period when oil prices dropped from USD 110 to USD 30 per barrel. We cut our investment programme in the areas where it did not directly affect our workers and development prospects. We adapted quickly and did not drastically decrease oil production. We did these things fairly smoothly and flexibly. I hope we will continue to do so in the future. This permits us to maintain our production volume at 110–120 million tonnes per annum. (Neft Rossii 2016)

Alekperov has always had strong (and sometimes wrong) opinions about future oil prices. For example, in a 2006 interview with the newspaper *Vedomosti*, when asked about his forecast of world oil prices, Alekperov remarked:

> In general, the market now is saturated, and oil, production and consumption are balanced. Therefore, as long as there are no political, climatic, or technogenic cataclysms, a major rise of the oil price is not anticipated. Its decline is possible, but it also would not be significant because consumption in China and India is increasing at really impressive rates. (Vedomosti 2006b)

During the 2009 crisis, Alekperov said that there were no fundamental reasons for oil prices to be below USD 40 and above USD 70 (Alekperov 2009). At the time, he was confident that oil prices would continue to grow. Calculating investments for 2009, LUKOIL proceeded from three scenarios: optimistic (USD 80 per barrel), acceptable (USD 65) and pessimistic (USD 45). LUKOIL's investment programme of USD 7.3 billion was based on a price range of USD 45–50 per barrel (Vedomosti 2009e).

In 2009, Alekperov was convinced that the oil price would continue to increase and would not drop below USD 40. By contrast, in 2010, Leonid Fedun expressed his view that the energy market was on the brink of a revolution due to shale gas production in the United States and rising oil production in Iraq, and that oil prices would never be high again.

LUKOIL's 2030 market forecast contains several scenarios, two of which are titled Concord and Volatility, and LUKOIL analysts believe that 'the average level of oil prices up to 2030 will be close to the level of USD 80 per barrel in constant prices' (LUKOIL 2018i).

Sanctions

LUKOIL was mainly affected by the sectoral sanctions targeting the Arctic, shale and deep-water oil (the company was not subject to the financial sanctions). As mentioned, it had to freeze its joint venture with Total on the development of the Bazhenov shale formation. Its deep-water projects in the Caspian Sea were also affected. TsentrCaspneftegaz, a joint venture between Gazprom and LUKOIL and the licence holder for Tsentralnaya deep-water structure in the Russian sector of the Caspian Sea, had to postpone drilling until sanctions were lifted. The joint venture could not obtain a drilling rig from foreign companies; therefore, the partners decided to suspend the project for the time being (TASS 2015).

In the context of sanctions, LUKOIL began to think about selling its oil-trading entity Litasco, established in 2000 in Switzerland, because of high maintenance costs. Vagit Alekperov stated that 'I would not say that it is connected with the sanctions, although it is quite difficult for our company to raise external financing while trading transactions require huge turnover capital' (Kommersant 2017d).

LUKOIL was the first Russian oil company to discern the danger of the 2017 US sanctions for foreign assets held by Russian oil companies. Leonid Fedun admitted that LUKOIL was interested in overseas projects; however, the threat of new American sanctions held back the company's acquisitions. 'Now there is a potential threat connected with the new package of sanctions adopted by the United States that limit the possibilities of Russian oil companies to own more than 33% in the project. This constrains us very strongly in several acquisitions' (Kommersant 2017e).

In 2017, a Sberbank CIB analyst summed up LUKOIL's stance on the political and economic situation very clearly: 'the management [of LUKOIL] sobered up when sanctions were placed on Russia, which coincided with the drop in the oil price' (Sberbank CIB 2017, p. 27).

Climate Policy and Changing Demand for Fossil Fuels

LUKOIL appears to take climate change seriously. Since 2013, the company has participated in the international Carbon Disclosure Project, which requires companies to provide information on their GHG emissions. In 2016 and 2018, based on its results, LUKOIL was awarded a climate score of D (with A being the highest and F the lowest).

By comparison, BP, ENI, Statoil (now Equinor) and Total all achieved A- (A minus). However, most participating Russian oil and gas companies achieved a score of F, except for Gazprom, LUKOIL and Novatek, which achieved C, D and D– respectively. Since 2015, LUKOIL has been estimating direct GHG emissions of its subsidiaries using the methodology of the Russian Ministry of Natural Resources (LUKOIL 2016b, p. 67). In 2015–16, LUKOIL worked on developing a corporate system for measuring and managing GHG emissions. The total volume of LUKOIL's direct GHG emissions in Russia amounted to 31 million tonnes of CO_2 equivalent (LUKOIL 2016b, p. 49).

One of LUKOIL's measures to reduce the climate impact of its activities is increased utilization of APG. It was one of the first Russian oil companies to start reducing gas flaring, even before the state required this. LUKOIL approved the Programme of Rational Utilization of APG by Organizations of the LUKOIL Group for 2016–18.

In 2016, LUKOIL built and upgraded 42 facilities for the utilization of APG. The main efforts were concentrated in Timan-Pechora and the North Caspian. The greatest increment of APG utilization was achieved in Timan-Pechora, where the volume of utilization increased by 267 million cubic metres (16%) from its 2015 level. However, in the North Caspian, due to the commissioning of the Filanovskogo field and gas production growth, APG utilization declined. In 2016, LUKOIL achieved an APG utilization level of 92% across the company and 97% in Western Siberia (LUKOIL 2016b, p. 23). It aimed to achieve 95% APG utilization in its Russian subsidiaries by 2018 (LUKOIL 2016b, p. 47). It is noteworthy that the company exceeded this target: in 2018, its level of APG utilization in Russia reached 97.4% (LUKOIL 2019b).

In addition, LUKOIL is, in effect, the only vertically integrated oil company in Russia that has shown an interest in renewables. It is noteworthy that its 2030 energy outlook states that 'oil and renewables are not antagonists' (LUKOIL 2018i). In 2016, LUKOIL had four hydropower stations in Russia with a total capacity of 296 MW and an output of 756 million kW/h. Its total capacity of renewable energy sources, including hydropower, was 390 MW in 2016. LUKOIL first ventured beyond hydropower with pilot projects in 2009 to equip three fuel stations in Serbia and Russia with photovoltaic panels. A renewable energy competence centre to serve LUKOIL units was established based on LUKOIL-Ecoenergo (LUKOIL 2016b, p. 20).

In the 2010s, LUKOIL focused its attention on wind power. In 2013, its 50:50 joint venture with the Italian company ERG, Lukerg Renew, bought four Bulgarian companies that specialized in wind power generation (Globo Energy, Mark 1, Mark 2 and UP Bulgaria 4). Lukerg was established by LUKOIL and Erg SpA in 2011 to jointly invest in renewables in Bulgaria and Romania. The joint venture owns a 40 MW wind farm in Bulgaria and a 84 MW wind power station in Romania (Kommersant 2013b). LUKOIL also has two solar power stations at its refineries in Romania with a total capacity of 9 MW and in Bulgaria with a capacity of 1.3 MW. These stations supply electricity to local networks.

Vagit Alekperov made the following observation:

> We have both traditional and solar power generation in Romania and Bulgaria. We entered this segment in Europe to gain experience. Also, because the EU governments promote these projects, they are quite profitable. We wanted to apply our experience in Russia. We decided to install solar modules in the Volgograd refinery, where we launched a 12-MW project. (Neft Rossii 2016)

Vagit Alekperov also recently founded a chair of renewable energy in the Gubkin Oil and Gas Academy led by one of LUKOIL's senior managers.

CONCLUSION

LUKOIL, the vanguard of change in the Russian oil sector during the 1990s, demonstrated considerable flexibility in adapting to internal changes in Russia during the 2000s and 2010s and swiftly adjusting its strategies. It also demonstrated considerable awareness of external changes, such as low oil prices, climate change and the rise of shale oil, and the ability to successfully cope with them. The company was at the forefront of reform in the industry, driving privatization in the 1990s and technological innovation pertaining to shale oil in the 2000s and 2010s. It responded flexibly to the challenges posed by oil price volatility and sanctions.

While LUKOIL enjoys close ties with the Russian authorities and appears committed to paying taxes, the company has also been keen to explore international opportunities through partnerships with foreign companies as well as to develop upstream and downstream business abroad both in the post-Soviet states and further afield. Sanctions have hampered its ability to reach its full potential in foreign markets. Nonetheless, it has managed to develop important foreign oil and gas fields and has established extensive fuel distribution networks. LUKOIL also prides itself on its support for technological innovation in several areas, including hard-to-recover reserves and renewable energy.

The company has paid some attention to climate change and GHG emissions by measuring its climate change impact, the utilization of associated gas rather than flaring it, and developing new climate change mitigation technologies.

LUKOIL seeks to maintain its reputation as a good corporate citizen who engages in various social investment activities across the areas where it operates. In Russia, this is done largely via agreements negotiated with local authorities. Successive sustainability reports indicate LUKOIL's growing awareness of the need to strengthen its communication with the general public concerning environmental protection and the well-being of local communities in its regions of operations. The company has, however, faced challenges in maintaining its social licence to operate in areas particularly affected by oil spills, notably the Komi Republic where local communities actively protested and succeeded in halting LUKOIL's operations – at least for a while. This event may have also convinced the company's management to bring community concerns to the fore in its public relations announcements.

Transparency is important to LUKOIL, and it has committed itself to several international and domestic initiatives requiring transparency and reporting. It has performed relatively well in ranking exercises that have assessed the transparency and sustainability performance of oil companies. More importantly, these indicators indicate improvement over time.

Table 3.3   LUKOIL's overseas projects

	Project/region	Partners	Timeline	Further details
Uzbekistan	Kandym-Hauzak-Shady gas project	LUKOIL (90%); Uzbekneftegaz (10%)	2004: a 36-year PSA is signed 2007: first gas from Hauzak 2011: first gas from Western Shady 2017: start of construction, Kandym gas-processing plant (gas capacity: 8 billion cubic metres)	The gas produced by the project was to be sold to Gazprom at the Uzbek-Kazakh border (Neft Rossii 2007). LUKOIL also sells a proportion to CNPC.
	South-West Gissar		2007: a 36-year PSA is signed 2008: LUKOIL joins the project 2013: full-scale implementation begins	LUKOIL expects to produce 18 billion cubic metres of gas from this project and the Kandym-Hauzak-Shady project by 2020 (Kommersant 2017f).
	Aral Sea	A consortium of Uzbekneftegaz, LUKOIL Overseas, Petronas Carigali Overseas, KNOC and CNPC, each owning 20% (Vedomosti 2005d)	2005: the consortium is set up 2010s: Petronas and KNOC withdraw from the consortium 2017: LUKOIL, CNPC and Uzbekneftegaz decide to wind up the consortium	Several small fields discovered, including Zapadniy Aral with 9 billion cubic metres of recoverable gas reserves; however, the project has not been taken forward.

	Project/region	Partners	Timeline	Further details
Kazakhstan	Kumkol	Turgai Petroleum (operator) includes LUKOIL (50%) and CNPC (50%)	1995: LUKOIL joins the development of northern Kumkol field 1996: a 25-year contract is signed for field development	In 2016, LUKOIL's share of production amounted to 0.3 million tonnes of oil and 26 million cubic metres of gas (LUKOIL 2018k).
	Karachaganak	Karachaganak Petroleum Operating includes co-operators BG and ENI (29.25% each) with Chevron (18%), LUKOIL (13.5%) and Kazmunaigaz (10%)	1997: a 40-year PSA is signed	Karachaganak field has reserves of 1.2 billion tonnes of oil and condensate and over 1.35 trillion cubic metres of gas. Karachaganak accounts for 45% of gas and 16% of liquids produced in Kazakhstan.
	Tengiz and Korolevskoye fields	TengizChevroil (operator) comprises Chevron (50%), ExxonMobil (25%), Kazmunaigaz (20%) and Lukarco (5%)	1993: a 40-year field development agreement is signed 1997: LUKOIL joins the project	The fields have recoverable reserves of oil, ranging from 750 million tonnes to 1.1 billion tonnes. In 2016, LUKOIL's share in production amounted to 1.4 million tonnes of oil and 0.5 billion cubic metres of gas.
Kazakhstan / Russia	Caspian Pipeline Consortium	Transneft (31%), Republic of Kazakhstan (20.75%), Chevron (15%), Lukarco (12.5%), ExxonMobil (7.5%), Rosneft-Shell (7.5%), BG (2%), ENI (2%) and Oryx Caspian Pipeline (1.75%)		This is the only private trunk pipeline in Russia and Kazakhstan. The pipeline is laid from Tengiz to Novorossiysk.

	Project/region	Partners	Timeline	Further details
Azerbaijan	Azeri-Chirag-Guneshli		1994: LUKOIL joins the project with 10% 2003: LUKOIL sells its stake to Japanese INPEX Southwest Caspian Sea Ltd for USD 1.3 billion (Neft i kapital 2003)	
	Shah-Deniz	BP (operator) (28.83%), TPAO (19%), Petronas (15.5%), SOCAR (10%), LUKOIL (10%), NICO (10%) and SGC (6.67%)	1996: a 40-year PSA is signed 2001: commercial reserves are found 2006: commercial production begins	Gas condensate fields located at depths of 700 metres; total reserves of 1.2 trillion cubic metres of gas and 240 million tonnes of condensate; in 2016, LUKOIL's share in production amounted to 0.9 billion cubic metres of gas and 0.2 million tonnes of condensate.
Venezuela	National Oil Consortium	Rosneft	2008: LUKOIL joins the consortium with a 20% stake 2013: LUKOIL sells its 20% share to Rosneft	LUKOIL left the project because it was more interested in projects where it could play a managerial role (Kommersant 2013c).
Colombia	Condor Block	Partnership with Ecopetrol; as the operator, LUKOIL had 70% (Vedomosti 2006c)	2002: LUKOIL gets involved in exploration activities 2012: LUKOIL decides to leave Colombia, having found no commercial reserves (Kommersant 2012f)	

	Project/region	Partners	Timeline	Further details
Mexico	Joint E&P activities	LUKOIL and Pemex	2014: LUKOIL signs an agreement with Pemex on joint E&P, the first foreign company to do so after Mexico's energy reform (Vedomosti 2014)	LUKOIL's priority is offshore projects in Mexico (Kommersant 2017g).
	Amatitlan onshore block	LUKOIL Europe (50%) and Marak Capital S.A.P.I. (50%)	2015: Lumex Holding B.V., a joint venture of LUKOIL Europe Holdings, and Marak Capital acquired Petrolera de Amatitlan S.A.P.I. from GPA Energy; each entity acquired a 50% interest and entered into a service contract with Pemex-Exploración y Producción	

	Project/region	Partners	Timeline	Further details
Iraq	West Kurna-2	*Original consortium:* LUKOIL (68.5%), Iraqi Ministry of Oil (25%), Mashinoimport (3.5%), Zarubezhneft (3.5%) *2009 consortium:* Iraq 25%, LUKOIL 63.75% and Statoil (now Equinor) 11.25% (Vedomosti 2009f) *Current consortium:* Iraqi South Oil Company and LUKOIL (75%), North Oil Company (25%) (Kommersant 2012g)	1997: PSA up to 2020 signed LUKOIL unable to work at the field due to UN sanctions 2002: Iraqi Ministry of Oil terminates LUKOIL's contract 2009: LUKOIL wins the tender for the development of the project together with Statoil (now Equinor) (not ConocoPhillips as had been expected) 2012: Statoil (now Equinor) withdraws from the project; LUKOIL buys out its stake 2016: LUKOIL intends to launch the second phase of West Kurna-2 in 2022	Discovered in 1973, West Kurna-2 has reserves of 1.8 billion tonnes of oil. The partners planned to invest some USD 5 billion in the project. In 2016, LUKOIL's share in West Kurna-2 oil production was 5 million tonnes. Maximum oil production could be 64.2 million tonnes when all three phases are implemented (Rambler 2014).
	Block 10	LUKOIL (60%), INPEX (Japan) (40%)	2012: partners win the fourth licensing round for the block (Kommersant 2012h) 2013–14: focus on removing land mines in the project area 2014–15: 2D seismic testing 2017: LUKOIL and INPEX successfully test the first exploration well Eridu-1, confirming the presence of a major petroleum field	

Project/region		Partners	Timeline	Further details
Egypt	Meleiya	ENI (operator) (76%), LUKOIL (24%)	1995: LUKOIL joins the project, one of its first overseas E&P ventures. Highly productive fields discovered in: 2007: North Nada and Gavaher 2010: Arcadia 2012: Emri Dip 2013: Rosa North 2014: Meleiya-West	In 2016, LUKOIL's share of oil production amounted to 182 000 tonnes.
	WEEM		1998: production starts. 2002: LUKOIL joins the project	In 2016, LUKOIL's share of production amounted to some 110 000 tonnes.
	WEEM Extension	LUKOIL (50%), Tharwa Petroleum (50%)	2009: LUKOIL joins the project	In 2016, LUKOIL's share of oil production amounted to 5000 tonnes.
Saudi Arabia	Block A, Rub Al-Hali petroleum basin	LUKSAR joint venture: LUKOIL (80%) and Saudi Aramco (20%)	2004: LUKSAR establishes and wins an international tender for the exploration and production of gas and condensate at Block A The Saudi government and LUKSAR sign a 40-year agreement for the development of the block In 2007, LUKOIL announces the discovery of commercial reserves of gas at Block A (Vedomosti 2007) 2016: LUKOIL withdraws from the project against a backdrop of falling oil prices (RBK 2016)	

	Project/region	Partners	Timeline	Further details
Cote d'Ivoire	CI-205 deep-water block, Gulf of Guinea	LUKOIL (63%)	2006: LUKOIL Overseas buys a 63% stake in a PSA 2016: LUKOIL pulls out due to low oil prices (Vedomosti 2006d)	LUKOIL had invested about USD 800 million but found no significant reserves.
Sierra Leone	SL-4B-10 offshore exploration block	Talisman Energy (operator), LUKOIL (25%)	2012: LUKOIL purchases a 25% stake from Talisman Energy	
	SL-5-11 offshore block (Kommersant 2012e)	LUKOIL (49%)	2011: LUKOIL joins the project as the operator; the participants are Nigerian Oranto (30%) and American PanAtlantic (21%) 2016: LUKOIL withdraws from the project (Kommersant 2016b)	
Ghana	Tano/Cape Three Points deep-water block		2014: LUKOIL joins the project	Seven hydrocarbon fields discovered within the block. LUKOIL adopts a 'wait-and-see' policy, hoping for higher oil prices.
Cameroon	Etinde deep-water project	New Age Ltd (operator) with 30%, LUKOIL (30%), EurOil Ltd (20%) and the Cameroon Societe Nationale des Hydrocarbures (20%)	2008: PSA signed 2014: LUKOIL joins the project	LUKOIL adopts a 'wait-and-see' policy, hoping for higher oil prices.
Nigeria	OML-140 deep-water project	Chevron (operator) (27%), NNPC (30%), ONG (25%), LUKOIL (18%)	2009: Development licence for the block is issued 2014: LUKOIL joins the project	The Nsiko field is discovered, and several promising structures are identified. LUKOIL adopts a 'wait-and-see' policy, hoping for higher oil prices.

Project/region	Partners	Timeline	Further details
Vietnam	Hanoi Trough-2	LUKOIL Overseas	2011: LUKOIL Overseas buys a 50% stake in the block from Quad Energy S.A. (Kommersant 2011b)	
Norway	Licensing plot PL 719 in the Barents Sea (two blocks 7321/8 and 7321/9)	Centrica Resources with 50% is operator, LUKOIL has 30% and North Energy 20%	2013: LUKOIL wins in the twenty-second licensing round	
	Licensing plot PL 708 (two blocks 7130/4 и 7130/7)	Project participants are Lundin Norway (operator) with 40%, LUKOIL (20%), Edison (20%), Pure Energy (10%) and Lime Petroleum (10%)	2013: LUKOIL wins licensing plot; 2015: a minor gas field is discovered (Kommersant 2013a); 2016: LUKOIL considers withdrawing from block 708 as field appears quite small, and LUKOIL is dissatisfied with the economics of the project (Kommersant 2013a)	
Romania	Trident offshore block, Black Sea Rapsodia block	LUKOIL (operator) (72%) in both projects, PanAtlantic (18%) and Romanian Romgaz (10%)	2011: LUKOIL wins the right to develop Rapsodia and Trident blocks; 2014–15: partners discover Lira field 2016: LUKOIL withdraws from Rapsodia block; 2017: LUKOIL plans to drill the third well in Trident block	

NOTE

1. The Imilorsko-Istochniy licensing area includes Imilorskoye, Zapadno-Imilorskoye and Istochnoye fields and is located in the Surgut region of Khanty-Mansi Autonomous District. In 2014, LUKOIL launched Imilorskoye field, one of the largest new fields of Western Siberia.

REFERENCES

Alekperov, V. (2009), 'Net fundamentalhykh prichin dlya tsen na neft nizhe $40 i vyshe $70', *Vedomosti*, 5 June.

Alekperov, V. (2011), *Neft Rossii: Proshloye, Nastoyaschee i Buduschee*, Moscow: Kreativnaya Ekonomika.

EITI (2018a), 'The global standard for the good governance of oil, gas and mineral resources', accessed 12 October 2018 at https://eiti.org/.

EITI (2018b), 'Companies', accessed 12 October 2018 at https://eiti.org/supporters/companies.

Energetika i Promyshlennost Rossii (2008), 'LUKOIL popal v italiyanskuyu pererabotku', accessed 12 October 2018 at https://www.eprussia.ru/pressa/articles/11218.htm.

Energyland (2008), 'LUKOIL vybral Rossiyu', accessed 12 October 2018 at http://www.energyland.info/analitic-show-5229.

EURACTIV (2018), 'LUKOIL CEO: Russian oil business prospers in EU', accessed 12 October 2018 at https://www.euractiv.com/section/energy/interview/lukoil-ceo-russian-oil-business-prospers-in-eu/.

Forbes (2009), 'Ostorozhno LUKOIL', accessed 12 October 2018 at http://www.Forbes.ru/ekonomika/kompanii/17235-ostorozhno-C2%ABlukoil%C2%BB.

Gaiduk, I. and O. Lukin (2006), 'LUKOIL moving into Europe', accessed 12 October 2018 at www.wtexecutive.com/cms/content.jsp?id=com.tms.cms.article.article_rpi_insight_LUKOILEurope.

Global Reporting (2019), 'About', accessed 22 January 2019 at https://www.globalreporting.org/Pages/default.aspx.

Gorst, I. (2007), 'LUKOIL: Russia's largest oil company' written for 'The changing role of national oil companies in international energy market', James A. Baker III Institute for Public Policy Report.

Gorst, I. and N. Poussenkova (1998), 'Petroleum ambassadors of Russia' written for 'Unlocking the assets: Energy and the future of Central Asia and the Caucasus', James A. Baker III Institute for Public Policy Report.

Greenpeace (2014), 'Komi-izhemtcy obyavili LUKOIL personoy non-grata', accessed 14 October 2018 at http://www.greenpeace.org/russia/ru/news/blogs/green-planet/-/blog/48716/.

GRI (2018), 'GRI Standards', accessed 12 October 2018 at https://www.globalreporting.org/Pages/default.aspx.

Hudson, J. (1996), *Russian Oil, Vertically Integrated Companies*, Volume 3 of Report on Russian Oil, London: Salomon Brothers.

Kelman, I., J.S. Loe, E.W. Rowe, E. Wilson, N. Poussenkova, E. Nikitina and D.B. Fjærtoft (2016), 'Local perceptions of corporate social responsibility for Arctic petroleum in the Barents Region', *Arctic Review*, **7** (2), 152–78.

Khrennikov, I. (2007), 'LUKOIL: Pro zapas', *Forbes*, accessed 11 October 2018 at http://www.Forbes.ru/Forbes/issue/2007-06/12377-lukoil-pro-zapas.

Kommersant (1996), 'Sobraniye aktsionerov LUKOYLa: Neftyanaya kompaniya vpolne dovolna soboy', accessed 22 January 2019 at https://www.kommersant.ru/doc/134111.

Kommersant (2011a), 'LUKOIL otbil litsenziyu', accessed 12 October 2018 at https://www.kommersant.ru/doc/1689728.

Kommersant (2011b), 'LUKOIL proburit Vietnamskiy shelf', accessed 14 October 2018 at https://www.kommersant.ru/doc/1633888.

Kommersant (2012a), 'LUKOIL rasplatilsya za oshibki geologov', accessed 12 October 2018 at https://www.kommersant.ru/doc/1884165.

Kommersant (2012b), 'Minprirody vozvraschayet Trebsa i Titova', accessed 12 October 2018 at https://www.kommersant.ru/doc/1901026.

Kommersant (2012c), 'LUKOIL zabral u Rossii samoye dorogoye', accessed 12 October 2018 at https://www.kommersant.ru/doc/2098523.

Kommersant (2012d), 'LUKOIL sokraschayet serbskuyu roznitsu', accessed 12 October 2018 at https://www.kommersant.ru/doc/2042339.

Kommersant (2012e), 'LUKOIL ukrepil Afrikanskiy blok', accessed 12 October 2018 at https://www.kommersant.ru/doc/2062229.

Kommersant (2012f), 'LUKOIL vypustil Kondora', accessed 14 October 2018 at https://www.kommersant.ru/doc/2088810.

Kommersant (2012g), 'LUKOIL ostaetsya v Irake odin', accessed 14 October 2018 at https://www.kommersant.ru/doc/1864335.

Kommersant (2012h), 'Neftyanoye peremiriye', accessed 14 October 2018 at https://www.kommersant.ru/doc/2041743.

Kommersant (2013a), 'Rossiya poluchila dolyu v Norvegii', accessed 12 October 2018 at https://www.kommersant.ru/doc/2210881https://www.kommersant.ru/doc/2210881.

Kommersant (2013b), 'LUKOIL dunul v Bolgariyu', accessed 14 October 2018 at https://www.kommersant.ru/gallery/2260031.

Kommersant (2013c), 'Venezuelu ostavyat goskompaniyam', accessed 14 October 2018 at http://enewz.ru/news/2001-venesuelu-ostavyat-goskompaniyam.html.

Kommersant (2016a), 'LUKOIL zasomnevalsya v Norvegii', accessed 12 October 2018 at https://www.kommersant.ru/doc/3150349.

Kommersant (2016b), 'LUKOIL razocharovalsya v Afrike', accessed 14 October 2018 at https://www.kommersant.ru/doc/2895093.

Kommersant (2017a), 'Kompaniy v Rossii vse menshe i menshe, i vse khuzhe i khuzhe aktivy, kotorye u nikh ostayutsya', accessed 11 October 2018 at https://www.kommersant.ru /doc/3508459.

Kommersant (2017b), 'LUKOIL nashel mesto na shelfe Rumynii', accessed 12 October 2018 at https://www.kommersant.ru/doc/3344753.

Kommersant (2017c), 'LUKOIL zatailsya na shelfe Afriki', accessed 12 October 2018 at https://www.kommersant.ru/doc/3325601.

Kommersant (2017d), 'LUKOIL distantsiruetsya ot svoego treidera', accessed 14 October 2018 at https://www.kommersant.ru/doc/3397679.

Kommersant (2017e), 'Seichas suschestvuet potentsialnaya ugroza, svyazannaya s novym paketom sanktsiy', accessed 14 October 2018 at https://www.kommersant .ru/doc/3416333.

Kommersant (2017f), 'LUKOIL vybralsya iz Arala', accessed 14 October 2018 at https://www.kommersant.ru/doc/3219937.

Kommersant (2017g), 'LUKOIL nashel partnera v Meksike', accessed 14 October 2018 at https://www.kommersant.ru/doc/3188639.

London Stock Exchange (2018), 'Rules and regulations', accessed 12 October 2018 at https://www.londonstockexchange.com/companies-and-advisors/main-market/rules/regulations.htm.

LUKOIL (2012), *Sustainability Report 2011–2012*, accessed 17 April 2019 at http://www.lukoil.com/InvestorAndShareholderCenter/ReportsAndPresentations/SustainabilityReport/SustainabilityReportArchive2003-2008.

LUKOIL (2016a), *Annual Report 2016*, accessed 12 October 2018 at http://www.lukoil.com/InvestorAndShareholderCenter/ReportsAndPresentations/FinancialReports.

LUKOIL (2016b), *Sustainable Development Report, 2015–2016*, accessed 12 October 2018 at http://www.lukoil.com/InvestorAndShareholderCenter/ReportsAndPresentations/SustainabilityReport.

LUKOIL (2017), *Making Opportunities Reality: LUKOIL Group Sustainability Report 2017*, accessed 21 April 2020 at https://csr2017.lukoil.com/download/full-reports/csr_en_annual-report_pages.pdf.

LUKOIL (2018a), 'LUKOIL's assets around the world & LUKOIL's assets in Russia', accessed 11 October 2018 at http://www.lukoil.ru/Business.

LUKOIL (2018b), 'Oil refining', accessed 11 October 2018 at http://www.lukoil.ru/Business/Downstream/OilRefining.

LUKOIL (2018c), 'Corporate profile', accessed 11 October 2018 at http://www.lukoil.Ru/Company/Corporateprofile.

LUKOIL (2018d), 'Growth projects', accessed 12 October 2018 at http://www.lukoil.ru/Business/Upstream/KeyProjects.

LUKOIL (2018e), 'Gas processing', accessed 12 October 2018 at http://www.lukoil.ru/Business/Downstream/GasProcessing.

LUKOIL (2018f), 'Petrochemicals', accessed 12 October 2018 at http://www.lukoil.ru/Business/Downstream/Petrochemicals.

LUKOIL (2018g), 'History. 2000', accessed 12 October 2018 at http://www.lukoil.ru/Company/history/History2000.

LUKOIL (2018h), 'LUKOIL charity fund', accessed 12 October 2018 at http://www.lukoil.com/Responsibility/SocialInvestment/SocialInitiatives/LUKOILCharityFund.

LUKOIL (2018i), 'Osnovnyye tendentcii razvitiya mirovogo rynka nefti do 2030 goda', accessed 14 October 2018 at http://www.lukoil.ru/FileSystem/PressCenter/84523.pdf.

LUKOIL (2018j), 'Sbyt', accessed 14 October 2018 at http://www.lukoil.ru/InvestorAndShareholderCenter/IrTool/InteractiveAnalysis/interactive-analysis-neww12?wid=widasul-rUAlkSyqp12NMVIdg.

LUKOIL (2018k), 'Zarubezhnyye proekty', accessed 14 October 2018 at http://www.lukoil.ru/Business/Upstream/Overseas.

LUKOIL (2018l), *Sustainable Development Report, 2018*, accessed 3 December 2019 at http://www.lukoil.com/InvestorAndShareholderCenter/ReportsAndPresentations/Sus tainabilityReport.

LUKOIL (2019a), 'Social Code of PJSC LUKOIL', accessed 22 January 2019 at http://www.lukoil.com/Responsibility/SocialPartnership/SocialCodeofPJSCLUKOIL/socialcodehtmlversion.

LUKOIL (2019b), 'Spravochnik Analitika', accessed 3 December 2019 at http://www.extraowa.lukoil.com.

Neft i kapital (2002), 'LUKOIL okonchatelno utratil vozmozhnost uchastvovat v konkurse po Gdanskomu NPZ', accessed 12 October 2018 at https://oilcapital.ru/news/markets/21-10-2002/lukoyl-okonchatelno-utratil-vozmozhnost-uchastvovat-v-konkurse-po-gdanskomu-npz.

Neft i kapital (2003), 'LUKOIL zavershayet prodazhu svoey doli v proekte Azeri-Chirag-Gyuneshli INPEKSu', accessed 12 October 2018 at https://oilcapital.ru/news/markets/28-04-2003/lukoyl-zavershaet-prodazhu-svoey-doli-v-proekte-azeri-chirag-gyuneshli-inpeksu.

Neft Rossii (2007), 'LUKOIL sbezhal ot Gazproma', accessed 14 October 2018 at http://www.oilru.com/news/59489/.

Neft Rossii (2016), 'Period "odin na odin" s gosudarstvom my uzhe prohodili', accessed 12 October 2018 at http://www.oilru.com/news/533114/.

Neftegaz.ru (2000), 'O finansovoy otchetnosti gruppy LUKOYL', accessed 22 January 2019 at https://neftegaz.ru/news/view/1321-O-FINANSOVOY-OTChETNOSTI-GRUPPY-LUKOYL.

Neftegaz.ru (2007), 'LUKOIL – v prezhnih granitsah', accessed 12 October 2018 at https://neftegaz.ru/press/view/3387.

Neftegaz.ru (2010), 'Moya missiya esche ne zakonchena', accessed 14 October 2018 at https://neftegaz.ru/news/view/96635-Alekperov-missiya-ne-zakonchena.

Neftegaz.ru (2017), 'RITEK prodoljayet vnedryat tekhnologiyu termogazovogo vozdeystviya na plasty Bazhenovskoy svity v HMAO', accessed 14 October 2018 at https://neftegaz.ru/news/view/159803-RITEK-prodolzhaet-vnedryat-tehnologiyu-termogazovogo-vozdeystviya-na-plasty-bazhenovskoy-svity-v-HMAO.

Neftrossii.ru (2014), 'Vagit Alekperov nagrazhdon ordenom "za zaslugi pered otechestvom" ii stepeni', accessed 22 January 2019 at https://neftrossii.ru/content/vagit-alekperov-nagrazhdyon-ordenom-za-zaslugi-pered-otechestvom-ii-stepeni.

Newsru.com (2002), 'LUKOIL obyavil o prodazhe svoyey doli v Azerbaidzhanskom proekte', accessed 12 October 2018 at https://www.newsru.com/finance/18nov2002/lukoil3.html.

Oilcareer.ru (2009), 'LUKOIL sokrashchayet personal?', accessed 14 October 2018 at http://www.oilcareer.ru/news/lukoil_sokrashhaet_personal/2009-10-28-162.

Oilgas.com (2017), 'RITEK pristupil k ispytaniyam tekhnologii termogazovogo vozdeystviya na plasty bazhenovskoy svity', accessed 14 October 2018 at http://oilgascom.com/%E2%80%A2-%E2%80%8Britek-pristupil-k-ispytaniyam-texnologii-termogazovogo-vozdejstviya-na-plasty-bazhenovskoj-svity/.

Poussenkova, N. (2010), 'Dvadtsat let, kotoryye potryasli mir', *Istoryya Novoy Rossyy*, accessed 12 October 2018 at http://www.ru-90.ru/node/1319.

Rambler (2014), 'Alekperov: LUKOIL verit v stabilnyy Irak', accessed 14 October 2018 at https://finance.rambler.ru/economics/25597462-alekperov-lukoyl-verit-v-stabilnyy-irak/.

RBK (2016), 'LUKOIL zayavil ob uhode iz Saudovskoy Aravii', accessed 14 October 2018 at https://www.rbc.ru/rbcfreenews/5762802a9a794724eff78999.

Reuters (2014), 'Russia's LUKOIL to drill for tight gas in Saudi desert', accessed 14 October 2018 at https://www.reuters.com/article/saudi-lukoil-gas/russias-lukoil-to-drill-for-tight-gas-in-saudi-desert-idUSL6N0O034G20140515.

Ria Novosti (2002), 'LUKOIL vykhodit iz proyekta Azeri-Chirag-Gyuneshli', accessed 12 October 2018 at https://ria.ru/economy/20021220/286344.html.

Russian Union of Industrialists and Entrepreneurs (2019), 'Social Charter of the Russian business', accessed 22 January 2019 at http://eng.rspp.ru/simplepage/860.

Sberbank CIB (2017), 'Russian oil and gas tomorrow is a distant memory', accessed 14 October 2018 at http://neftianka.ru/wp-content/uploads/2017/11/2 _5467772892869558529.pdf.pdf.

TASS (2015), 'Total pokroyet LUKOILu zatraty na razrabotku Bazhenovskoy svity pri vozvrashenii v proyekt', accessed 14 October 2018 at https://tass.ru/tek/2555532.

The Barrel (2015), 'Private companies struggle for Russian offshore opportunities: Regulation and environment', accessed 12 October 2018 at http://blogs.platts.com/ 2015/04/13/private-companies-offshore-russia-oil/.

Transparency International (2011), 'Promoting revenue transparency: 2011 report on oil and gas companies', accessed 12 October 2018 at https://www.transparency.org/ whatwedo/publication/promoting_revenue_transparency_2011_report_on_oil_and _gas_companies.

United Nations Global Compact (2019), 'Russia', accessed 22 January 2019 at https:// www.unglobalcompact.org/engage-locally/europe/russia.

Vedomosti (2005a), 'LUKOIL postavit benzin v SShA', accessed 12 October 2018 at https://neftegaz.ru/press/view/1836.

Vedomosti (2005b), 'LUKOIL potratil 2 milliarda', accessed 22 January 2019 at https://neftegaz.ru/press/view/1843.

Vedomosti (2005c), 'Alekperov dogovorilsya s Nazarbayevym', accessed 12 October 2018 at https://www.pressreader.com/russia/vedomosti/20051024/ 281745559782989.

Vedomosti (2005d), 'LUKOIL idyet v Uzbekistan', accessed 22 January 2019 at https://www.vedomosti.ru/newspaper/articles/2005/09/09/lukojl-idet-v-uzbekistan.

Vedomosti (2006a), 'LUKOIL stanet partnerom Petrol', accessed 22 January 2019 at https://www.eprussia.ru/pressa/articles/3866.htm.

Vedomosti (2006b), 'U kitaitsev bolshe vozmozhnostei, chem u nas', accessed 12 October 2018 at https://www.gubkin.ru/news/detail_ajax.php?ID=1633&sphrase_id =4517116#_Toc144772934.

Vedomosti (2006c), 'LUKOILu ponravilos v Kolumbii', accessed 22 January 2019 at https://www.vedomosti.ru/newspaper/articles/2006/04/11/lukojlu-ponravilos-v -kolumbii.

Vedomosti (2006d), 'LUKOIL dobralsya do zapadnoi Afriki', accessed 14 October 2018 at https://www.pressreader.com/russia/vedomosti/20060717/textview.

Vedomosti (2007), 'Gaz iz serdtsa OPEC', accessed 14 October 2018 at https://www .pressreader.com/russia/vedomosti/20070213/281878703906240.

Vedomosti (2008a), 'Ne nashli ni nefti, ni gaza', accessed 12 October 2018 at https:// www.vedomosti.ru/library/articles/2008/09/03/riskovye-francuzy.

Vedomosti (2008b), 'Gotovy k dobyche', accessed 22 January 2019 at https://www .eprussia.ru/pressa/articles/11185.htm.

Vedomosti (2008c), 'Don Alekperov', accessed 12 October 2018 at http://www .zagolovki.ru/article/25Jun2008/don.

Vedomosti (2009a), 'Opyat sukho', accessed 22 January 2019 at https://www.vedomosti .ru/newspaper/articles/2009/01/13/opyat-suho.

Vedomosti (2009b), 'BP ushla iz KTK', accessed 12 October 2018 at https://www .vedomosti.ru/newspaper/articles/2009/12/14/vr-ushla-iz-ktk.

Vedomosti (2009c), 'Ot Sitsilii do Gollandii', accessed 12 October 2018 at https://www .pressreader.com/russia/vedomosti/20090622/281646776119197.

Vedomosti (2009d), 'Oboshel Exxon', accessed 14 October 2018 at http://ukrrudprom .ua/digest/Oboshel_Exxon.html?print.

Vedomosti (2009e), 'Skolko stoit neft?', accessed 14 October 2018 at https://www
.pressreader.com/russia/vedomosti/20090605/281676840879096.

Vedomosti (2009f), 'LUKOIL vozvraschayetsya v Iraq', accessed 18 February 2019
at https://www.vedomosti.ru/business/articles/2009/12/14/lukojl-vozvraschaetsya-v
-irak%20.

Vedomosti (2010a), 'Patrioty iz LUKOILa', accessed 12 October 2018 at http://www
.energyland.info/analitic-show-51848.

Vedomosti (2010b), 'Postoronnim vhod zapreschen', accessed 12 October 2018 at
http://stocks.investfunds.ru/news/11114/.

Vedomosti (2010c), 'Otstupniye dlya LUKOILa', accessed 12 October 2018 at https://
www.pressreader.com/russia/vedomosti/20100823/281805690237144.

Vedomosti (2011), 'Amerike ne zalivat', accessed 12 October 2018 at http://www.case
-hr.com/novosti-rinka/35350.html.

Vedomosti (2012a), 'Rosnedra prepodnesli neftyannikam "vnezapniy syurpriz"',
accessed 12 October 2018 at https://www.vedomosti.ru/business/articles/2012/05/
22/partner stvo_vo_imya_nefti.

Vedomosti (2012b), 'LUKOIL peredumal idti na shelf s Rosneftiyu', accessed
12 October 2018 at https://www.vedomosti.ru/business/articles/2012/04/27/lukojl
_peredumal.

Vedomosti (2012c), 'LUKOIL zapravit Benilux', accessed 14 October 2018 at https://
www.vedomosti.ru/business/articles/2012/04/13/lukojl_zapravit_benilyuks.

Vedomosti (2013a), 'Razve ya pozhozh na cheloveka, kotoryi chto-to prodaet?',
accessed 21 April 2020 at http://enkorr.com.ua/a/interview/Vagit_Alekperov_Razve
_ya_pohozh_na_ cheloveka_kotoriy_chto-to_prodaet/209806.

Vedomosti (2013b), 'LUKOIL vysaditsya na Sitsilii', accessed 12 October 2018 at
http://stocks.investfunds.ru/news/49591/.

Vedomosti (2013c), 'Avoski i valenki obyedinyayutsya', accessed 14 October 2018 at
https://www.pressreader.com/russia/vedomosti/20131011/textview.

Vedomosti (2014), 'LUKOIL idet v Meksiku', accessed 14 October 2018 at http://
stocks.investfunds.ru/news/55193/.

Vedomosti (2016a), 'Dohodnoye mesto LUKOILa', accessed 12 October 2018 at
https://www.vedomosti.ru/business/articles/2016/12/16/669889-dohodnim
-regionom-lukoila.

Vedomosti (2016b), 'Lishniy benzin', accessed 22 January 2019 at https://www
.vedomosti.ru/business/articles/2016/02/08/627725-lukoil-zapravok-baltii.

Vestifinance.ru (2014), 'LUKOIL i Total sozdali SP po Bazhenovskoy svite', accessed
14 October 2018 at https://www.vestifinance.ru/articles/43146.

4. Gazprom Neft: reformed rake

Gazprom Neft is the oil arm of Gazprom, which is the world's largest gas company by any measure. Gazprom Neft was created on the basis of the oil company Sibneft, which was owned by the Russian oligarch Roman Abramovich, who sold it to Gazprom for a handsome price. The acquisition of Sibneft facilitated Gazprom's foray into the oil industry. A key question is whether the former Sibneft has managed to retain its unique identity, culture and strategy or has come to function more like a department of Gazprom.

Although Gazprom Neft is now a Gazprom subsidiary, it is also a major corporation in its own right and the country's third-largest oil producer and third-largest oil refiner. It is also the only company that produces significant amounts of oil on Russia's Arctic continental shelf, from the Prirazlomnoye field in the Pechora Sea.

With 95.68% of the shares, Gazprom fully controls Gazprom Neft; the remaining shares are traded on the open market (Gazprom Neft 2017c). Gazprom Neft's headquarters are in Saint Petersburg, the city to which Gazprom proper is also moving from Moscow.

The general directorship of Sibneft and then Gazprom Neft has been fluid. Abramovich was unwilling to take up the post of general director of Sibneft, although he ran the company to all intents and purposes. Sibneft had three different general directors in its first five years from 1995 to 2000; then the directorship changed twice again after Gazprom acquired Sibneft in 2005.

In 2017, Gazprom Neft surpassed Surgutneftegas and for the first time became the third-largest oil producer in Russia (Gazprom Neft 2017a). In 2018, its production of hydrocarbons increased by 3.5% from the 2017 level to 92.9 million tonnes of oil equivalent. Taking into account Gazprom Neft's shares in joint ventures, as of December 2018 its total reserves of hydrocarbons under the international 2P category (proved+probable) amounted to 2.84 billion tonnes of oil equivalent (Gazprom Neft 2018r).

Gazprom Neft oversees over 70 oil-producing, refining and marketing organizations in Russia, the former Soviet Union and other countries. It holds 50% stakes in several affiliated companies, namely Slavneft, Tomskneft, Salym Petroleum Development, Messoyakhaneftegaz, Northgaz and SeverEnergiya (Gazprom Neft 2017b). Its upstream operations are concentrated in Khanty-Mansi and Yamal-Nenets Autonomous Districts and Tomsk, Omsk and Orenburg regions. Its main refining facilities are located in Moscow,

Omsk, Yaroslavl, Belarus and Serbia. Gazprom Neft conducts exploration and production activities in Angola, Bosnia and Herzegovina, Hungary, Iraq, Romania, Serbia and Venezuela. Its network of fuel stations comprises 1859 units in Russia, the Commonwealth of Independent States (CIS) and Europe.

CORPORATE HISTORY

The 1990s: The Oligarchs' Playground

Sibneft was founded by special Presidential Decree 872 of 24 August 1995, on the basis of some of Rosneft's most precious assets. This contrasts with other Russian oil companies, which were established by standard government orders. At the time, Sibneft included Noyabrskneftegaz, the 22-million-tonne oil producer located in the Yamal-Nenets Autonomous District, the Omsk refinery, one of the largest and most modern in Russia, and the exploration company Noyabrskgeofisika and Omsknefteprodukt, an oil product distribution company. Investment analysts commented at the time that 'Sibneft has only four subsidiaries, but they are all top class' (Salomon Brothers 1996, p. 86).

Noyabrskneftegaz was, in effect, one of the youngest oil producers in Western Siberia, with a relatively low water-cut in its reservoirs, low field depletion rates and a high daily output from its wells. In the mid-1990s, despite financial difficulties, Noyabrskneftegaz was the only Russian oil company that managed to replace its reserves through exploration drilling. Moreover, Noyabrskneftegaz was fortunate to have reserves of sweet and light oil, which could be supplied as Siberian Light blend and guaranteed high-quality refined products (TRINFICO 1996).

The creation of Sibneft involved a scandal. Setting up the company required the support of two powerful businessmen who controlled key assets, Victor Gorodilov, the General Director of Noyabrskneftegaz, and Ivan Litskevich, who was the General Director of the Omsk refinery and a well-known advocate of the interests of the Omsk region. While Gorodilov supported the establishment of Sibneft, Litskevich opposed it. Decree 872 was signed only days after Ivan Litskevich had drowned while swimming in a river (TRINFICO 1996).

In 1995, Victor Gorodilov became the first General Director of Sibneft. Gorodilov was replaced in 1997 by Andrei Blokh, who had previously worked as the Deputy Director of Ecogazmotor and Deputy General Director of Petroltrans (two small, low-profile companies). Gorodilov's son Andrei Gorodilov continued to work in Sibneft for several years. Blokh was replaced two years later by Evgeniy Shvidler, a long-time business acquaintance of Abramovich and his partner in the oil-trading venture Runicom. Shvidler had joined Sibneft at the outset as Senior Vice President. He served as General Director until Sibneft's acquisition by Gazprom in 2005 (Knyazev 2010).

Sibneft has a murky past, and its history is closely connected with two of the most famous oligarchs of the post-Soviet period: Boris Berezovskiy and Roman Abramovich (Lysova 2005). The 'trial of the century' in the High Court of London in 2011 (see below) threw new light on what happened in the 1990s. It confirmed that Sibneft was bought under a corrupt scheme, as was the case with most Russian business in the 1990s. Boris Berezovskiy filed a lawsuit against Roman Abramovich in 2007 demanding USD 5.5 billion in compensation for damages that he claimed he suffered when he sold his stakes in Sibneft and Rusal, allegedly below the market value and under pressure and threats from Roman Abramovich. During the trial, it was revealed that in 1995 Boris Berezovskiy had convinced Boris Yeltsin to establish Sibneft and privatize it in such a way that Roman Abramovich would end up owning the company. Some experts suggest that Sibneft was expected to sponsor the ORT TV channel that Berezovskiy controlled then and that would have supported Yeltsin during the 1996 presidential elections (Zanina 2011).

In fact, Yeltsin ensured that Sibneft was privatized in phases; the bulk of the government stake was sold through the loans-for-shares auction. In December 1995, Neftyanaya Finansovaya Kompaniya (NFK), controlled by Boris Berezovskiy and Roman Abramovich, bought a 51% stake in Sibneft that was held by the government, paying USD 100 million (Neft i kapital 2004a). In January 1996, another block of shares (14.28%) was auctioned off, and it was understood that NFK was again the buyer (Rozhkova and Reznik 2009). In September and October 1996, 19% and 15% were sold at investment tenders for USD 45 million and USD 35.5 million, respectively. Finally, in December 1996, a 0.72% stake was sold at an auction. All these stakes were acquired by companies controlled by Abramovich and Berezovskiy (Rozhkova and Reznik 2009).

Ultimately, the Abramovich-Berezovskiy team gained control of 90% of Sibneft. However, the ownership structure of Sibneft was never transparent. When Sibneft was privatized, Abramovich was not even mentioned: Berezovskiy was considered to be the buyer of the company. It was only in December 1999 that Abramovich admitted in an interview with the business newspaper *Vedomosti* that he was the co-owner of Sibneft (Rozhkova and Reznik 2009). In 2011, Abramovich stated in the High Court of London that the rumours that Berezovskiy was a shareholder of Sibneft actually helped the company, given the political influence of Boris Berezovskiy at the time (Forbes 2011). Ultimately Abramovich won the lawsuit in London.

At the very beginning, in a foretaste of things to come much later, Sibneft was linked to Gazprom. In December 1996, Rem Vyakhirev, Head of the gas monopoly, was appointed chairman of Sibneft's board of directors; however, in September 1997, he declined re-election to the board because of his heavy workload at Gazprom (Neftegaz 1999).

Sibneft made its first attempt to merge with YUKOS in 1998; however, the resulting Yuksi 'super-company' existed for only six months. Experts widely blamed the psychological incompatibility of the individuals behind the two corporations (Berezovskiy, Abramovich and Mikhail Khodorkovskiy) for the failure of the merger (Neft i kapital 2004b).

In the second half of the 1990s, Sibneft was considered to be the most likely candidate to purchase Rosneft, which was slated for privatization (see the chapter on Rosneft). However, Sibneft's ambitions did not materialize, mainly because the privatization of Rosneft was postponed indefinitely.

In 1998, Sibneft was the third Russian oil company after LUKOIL and Surgutneftegas to consolidate shares as it began to swap shares in its subsidiaries for shares of the Sibneft holding. Sibneft used a unique approach to these share swaps: instead of the usually complicated process, the holding simply wrote letters to the shareholders of Noyabrskneftegaz and Noyabrskgeofizika offering to exchange shares with them. Analysts subsequently concluded that the proposed swap ratios were fair and justified (Savushkin 1998).

Sibneft was the first Russian oil company to issue Eurobonds (in September 1997, these were worth USD 150 million). It was also the first to publish consolidated financial statements under generally accepted accounting principles (GAAP) and conduct an international reserve audit (Neft i kapital 2004c). In 1997, the trade of Sibneft's shares began in the Russian Trading System (RTS), and in 1999, the company issued Level-I American depository receipts (ADRs), which were traded in the United States as well as on the Berlin and Frankfurt stock exchanges.

In the late 1990s, Elf Aquitaine sought to buy 5% of Yuksi; however, after the collapse of Yuksi, negotiations shifted to the purchase of some 12% of Sibneft. However, in September 1998, as the financial crisis hit and the oil price plummeted, Elf Aquitaine abandoned all its plans (Vinogradov 1998).

2000–05: The Rake's Progress

In the early 2000s, during the prolonged period of rising oil prices, Sibneft and YUKOS drove crude production growth in Russia. Sibneft was the favourite of investors because of its record high growth rates combined with low capital and operating costs. In 2001, its oil output grew by 20%, and in 2002 and 2003, it increased by 27% and 19%, respectively, while its lifting costs were only USD 1.7–1.8 per barrel (cf. USD 2.8–3 for LUKOIL and Tyumen Oil Company (TNK)). Sibneft was fortunate to have excellent reserves, particularly in terms of the high average daily output of wells (Figure 4.1). At the same time, its share of idle wells was over 50% in 2002.

Both Sibneft and YUKOS intensified production through the aggressive use of hydrofracking[1] and horizontal drilling, which they performed with the help of Schlumberger and Halliburton. Sibneft and YUKOS made significant investments in production, which were recouped through the rapid growth of oil output. Similarly to YUKOS, Sibneft introduced a new strategy focused on maximizing the current profitability of oil extraction with the ultimate goal of increasing the market value of the company (Neft i kapital 2004d). Sibneft and YUKOS were both strongly criticized for having applied enhanced oil recovery methods at the early stages of field development and for skimming the oilfields, since these approaches could allegedly damage reservoirs in the long term (Neft i kapital 2004d).

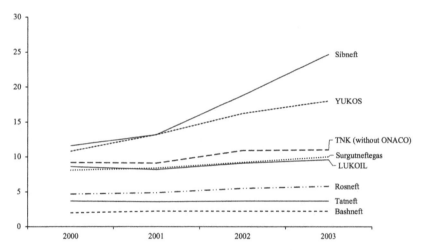

Figure 4.1   *Average daily output of operating wells of companies (tonnes/day), 2000–03*

Note: Rosneft (2003) without Severnaya Neft.
Source: Neft i kapital (2004d).

Other factors in Sibneft's growth included the commissioning of the major field Sugmutskoye in the Noyabrsk area of Yamal-Nenets Autonomous District in 2000 as well as the acquisitions that expanded the geography of its operations. In 2001, Sibneft and Chukotka Trading Company established Sibneft-Chukotka on a 50:50 basis. Chukotka Trading Company contributed to the authorized capital licences for the development of several fields (Lagunnoye, Telekaiskoye and Zapadno-Ozernoye). In 2004, Sibneft produced the first oil in Chukotka (Neft, gaz i fondovyy rynok 2008).

In December 2002, Sibneft bought 10.83% of Slavneft from Belarus for USD 207 million. Soon afterwards, an auction was held for the 74.95% of Slavneft owned by the Russian state. The winner was Investoil, created by Sibneft and TNK on a 50:50 basis. They paid USD 1.86 billion. A further 12.98% of Slavneft was already held by a trust company controlled by Sibneft and TNK (Neft, gaz i fondovyy rynok 2008). The new owners wanted to divide Slavneft; however, as they could not reach an agreement on the division or on whether one of the partners should buy out the other, they decided to divide Slavneft's oil, petroleum products and financial flows (Egorova 2006a).

During the early 2000s, Sibneft greatly strengthened its marketing arm mainly because it was considered a weakness that it had Omsknefteproduct as its sole petroleum product distributor. Thus, in 2000, Sibneft purchased a controlling interest in two product distribution companies in the Ural region, which enabled it to establish a strong foothold there. In 2001, it bought 78.4% of Tyumennefteproduct from TNK. In 2002 and 2003, Sibneft established oil product distribution subsidiaries in Krasnoyarsk and Saint Petersburg in order to develop its marketing networks in those regions (Neft, gaz i fondovyy rynok 2008).

During the 2000s, Sibneft fully demonstrated its traditional ability to turn conflicts over asset acquisition to its advantage. When ONACO – an 8-million-tonne-per-annum producer based in Orenburg – was to be privatized in 2000, a serious dispute arose between Sibneft and TNK over its assets. Sibneft joined forces with YUKOS and Stroitransgaz and began leading the alliance. It also bought 38% of ONACO's subsidiary, Orenburgneft, from YUKOS for USD 430 million in 2000. The starting price of the 85% state stake in ONACO was USD 425 million. During a lengthy round of bidding, the Sibneft-YUKOS-Stroitransgaz alliance offered USD 1 billion. TNK bid an additional USD 80 million and won. However, TNK was unhappy that Sibneft held 38% of Orenburgneft and began to negotiate with its rival to convince it to either sell or swap this stake. No agreement was reached until spring 2003 when Sibneft sold 38% of Orenburgneft and 1% of ONACO (which it had bought on the secondary market) to TNK for USD 825 million. By doing this, Sibneft made an impressive profit in a short space of time (Neft i kapital 2004e).

In the early 2000s, Sibneft consistently optimized tax payments through various schemes, resulting in one of the lowest profit tax rates in the sector. For instance, according to Sibneft's US GAAP statements, in 2003, it only paid 7% tax compared to the official Russian business rate of 35%. In February 2004, during the YUKOS case, the Ministry of Taxes and Duties made fiscal claims for the period 2000–01 against Sibneft amounting to USD 1 billion. The company reached a compromise by paying some USD 300 million to the state (NEWSru 2004). However, while saving on taxes, Sibneft always remained generous to its shareholders. In 2001, it paid over USD 1 billion in dividends, and in 2001, it paid USD 1 billion. In 2003, it paid no dividends because of its split from YUKOS; however, in 2004, the dividend payments amounted to USD 2.3 billion (Borisov 2005). According to some analysts, the shareholders pumped money out of Sibneft on the eve of its sale to Gazprom in 2005 (Egorova 2006b).

Sibneft has been involved in several high-profile corporate disputes, with the most infamous one being over the control of the Moscow refinery. In 2001, the Moscow city authorities had merged the refinery with Cyprus-based Sibir Energy, which was established by Shalva Chigirinskiy and his partner Igor Kesayev to create a vertically integrated company – the Moscow Oil and Gas Company (69% Moscow authorities; 31% Sibir Energy). Sibneft bought a block of the refinery shares from LUKOIL in 2001 (see the chapter on LUKOIL). In 2002, Sibneft and Tatneft, a fellow shareholder, launched a corporate war against the Moscow city authorities and Sibir Energy for the control of the assets. During the corporate hostilities, which lasted six years, the parties established parallel managing bodies and blocked oil deliveries to the refinery through court decisions. When Gazprom took over Sibneft in 2005, it sought to resolve the issue. However, in June 2006, representatives of Gazprom Neft were not admitted to the annual general meeting of the Moscow refinery, supposedly because of some technicalities related to the letter of authorization (Reznik and Tutushkin 2006). In 2007, peaceful negotiations began and in 2008 the shareholders agreed on the joint management of the refinery. The parties to the conflict signed a memorandum on cooperation pledging that they would manage the refinery on a parity basis; henceforth, representatives of Gazprom Neft joined the managing board of the Moscow refinery (Tutushkin 2008a).

Another controversy, also with Sibir Energy, broke out over Sibneft-Yugra, which had been established in 2000 on a 50:50 basis by Sibir Energy and Sibneft. Sibir Energy contributed 99.34% of Yugraneft to the joint venture, including its three oilfields in Khanty-Mansi Autonomous District. Sibneft became the operator of the joint venture with the obligation to provide funds for the development of the fields. In 2004, Sibir Energy's management discovered that Sibneft had diluted the stake of Sibir Energy in Sibneft-Yugra from 50% to 1% through two additional issues of shares approved at shareholder meetings. Sibir Energy stated that its stake in Yugraneft had been diluted without its permission and began to fight for the return of the asset. For its part, Sibneft claimed that all transactions had the approval of Sibir Energy's main shareholders (Tutushkin 2005). The court battles between Sibneft and Sibir Energy went on for a long time without a clear result (Reznik and Derbilova 2005a).

Yet another scandal with far-reaching political implications surrounded the second merger attempt of Sibneft and YUKOS in spring 2003. The resulting company would have been Russia's largest in terms of oil production and reserves and the world's second-largest private oil producer. Neither company denied that a significant stake in the new venture would eventually be offered to a foreign investor (Lenta.ru 2003). It was expected that Mikhail Khodorkovskiy would lead the company, while Evgeniy Schvidler would be the chairman of the board of directors. The representatives from YUKOS and Sibneft indicated that they had received political approval for the transaction (Landes and Nagaria 2003). By October 2003, YUKOS had obtained some 92% of Sibneft's shares, providing in return USD 3 billion and 26% of its own shares. However, soon after the deal was closed, Mikhail Khodorkovskiy was arrested. Millhouse Capital – a UK-registered company, established in 2001 to manage Roman Abramovich's assets and led by Evgeniy Shvidler – demanded the termination of the transaction. After lengthy court battles, Millhouse recovered 57.5% of Sibneft's shares in October 2004 and another 14.5% in July 2005. By September 2004, YUKOS learned that Abramovich did not intend to return the USD 3 billion and decided to initiate a lawsuit against Millhouse in the London Court of International Arbitration. In 2006, YUKOS was declared bankrupt by the government, and its bankruptcy manager, appointed by the Russian government, reached an amicable settlement with Millhouse concerning this lawsuit (Surzhenko and Reznik 2006).

2005 to the Present: Reform of the Rake, Life under Gazprom's Tutelage

A new stage in the life of Sibneft began after it was acquired by Gazprom. In September 2005, Gazprom and Millhouse Capital signed an agreement on the purchase of 72.6% of Sibneft for USD 13 billion. Earlier, Gazprom had bought 3.016% of the company that was held by Gazprombank (RBC 2005), culminating in a 75.6% share. YUKOS owned 20% of the company, and this block was acquired along with its other assets (Derbilova 2005).

Alexander Ryazanov, the deputy chairman of Gazprom's managing board, was appointed Sibneft's Head, replacing Evgeniy Shvidler (Reznik and Derbilova 2005b). Ryazanov admitted in an interview with Vedomosti that Gazprom had overpaid for Sibneft, allowing its former shareholders to skim the cream off the transaction (Reznik 2006c). German Gref, who was a member of Gazprom's board of directors at the time, was categorically against the deal because he was sure that the purchase price was inflated. Ultimately, however, he gave in and Abramovich became the richest Russian in 2005, according to *Forbes* (Rozhkova and Reznik 2009).

Gazprom soon discovered that its new asset had unexpected flaws. First, its reserves turned out to be smaller than anticipated. Petroleum consultants

De Golyer & MacNaughton concluded that by the end of 2005, Sibneft's total liquid reserves had decreased by 17%, while proved and producing reserves had declined by 46% (as compared to the results of an audit by Miller and Lents at the beginning of 2005). However, by 2006, proved and producing gas reserves had subsequently doubled (Reznik 2006d). Second, in 2005, Sibneft's oil production decreased for the first time in many years from 34 million tonnes per annum to 33 million tonnes per annum (Neftegazovaya vertikal 2006a). In 2006, it was expected that given the current level of investments, by 2010 its oil output could shrink to around 17 million tonnes (without taking into account Sibneft-Yugra and Sibneft's stake in Slavneft). Valeriy Nesterov, from the private investment bank Troika Dialog, concluded that the former owners had artificially inflated the company's value before its sale. Other analysts noted that Sibneft conducted the exploration and development of new fields without due care and attention and that the extraordinary rates of production growth achieved through aggressive hydrofracking were unsustainable. At the same time, Sibneft's extraction costs rose to USD 2.44 per barrel and were expected to continue growing (Reznik 2006a). They reached a record high level of USD 5.86 per barrel in 2008, making Sibneft's extraction costs some of the highest among Russian oil companies (Mazneva 2010a).

On 29 March 2006, Sibneft's board of directors decided to rename the company Gazprom Neft, and they re-registered it in Saint Petersburg (Reznik and Temkin 2006). Gazprom could not decide for a whole year what to do with its new asset. The monopoly's leading figures had divergent views on whether to leave the oil subsidiary as an independent company or incorporate it fully into the structure of Gazprom (Reznik 2006b). In September 2006, they decided that Gazprom Neft should remain a separate business unit within Gazprom, responsible for the development of its oilfields and the oil fringes of its gas fields (Tutushkin 2007).

In November 2006, Alexander Ryazanov was unexpectedly dismissed from his positions both in Gazprom and Gazprom Neft, despite his achievements as the Director of Gazprom Neft. Under his leadership, the company had stabilized oil production through large investments. Ryazanov had arranged for smaller dividends to be paid to shareholders and had allocated more funds for development. He insisted on transferring the Prirazlomnoye oilfield in the Pechora Sea from a Gazprom subsidiary to Gazprom Neft and arranged for the produced oil to be exported via an oil trader rather than Gazpromexport. Ryazanov's efforts to increase Gazprom Neft's independence might have annoyed two influential deputy chairmen of Gazprom: Alexander Medvedev, who insisted on Gazprom having a single export channel for all its production, and Alexander Ananenkov, who lobbied for the complete integration of the company into Gazprom's structure. Also, Ryazanov promoted Gazprom's purchase of a 20% stake of Sibneft from YUKOS. However, Rosneft coveted this 20% stake. Some experts believed that Ryazanov might have thus become a headache for the chairman of Rosneft's board of directors Igor Sechin (Neftegazovaya vertikal 2006b).

Alexander Dyukov, the President of petrochemicals company Sibur Holding and an old-time acquaintance of the chairman of Gazprom's management committee Alexei Miller, was appointed CEO of Gazprom Neft, replacing Ryazanov (Malkova and Reznik 2006). In April 2007, during the sale of YUKOS assets, the joint venture Enineftegaz, established by Gazprom, ENI and Enel, bought 20% of Gazprom Neft and some gas subsidiaries of YUKOS (Mazneva 2007).

Another scandal erupted during the sale of YUKOS' oil assets, which mostly concerned Rosneft (see the chapter on Rosneft). In May 2007, a Rosneft subsidiary bought several assets from YUKOS, including Tomskneft. Rosneft then announced that it would sell a 50% stake in Tomskneft to Vnesheconombank (reportedly working on behalf of Gazprom Neft). When Rosneft stated that the deal was closed, Vnesheconombank denied this (Petrachkova and Derbilova 2007). However, the scandal was hushed up, and in December 2007, Gazprom Neft directly acquired half of Tomskneft for USD 3.6 billion (Malkova and Mazneva 2007).

In the best traditions of Sibneft, Gazprom Neft continued to grow through acquisitions. In 2009, in fierce competition with TNK-BP, Gazprom Neft bought 16% of Sibir Energy, which permitted Gazprom Neft to consolidate 52% of the Moscow refinery indirectly. Gazprom Neft was also interested in Sibir Energy's participation in the Salym Petroleum Development – a joint venture with Shell created in 1996, which produced some 8 million tonnes of oil per annum (Baraulina et al. 2009). Gradually, Gazprom Neft increased its share in Sibir Energy to 80.37%, having spent USD 2.4 billion in total for this acquisition (Mazneva 2010b). In February 2011, the ten-year dispute over Sibir Energy and the Moscow refinery (see above) eventually ended, and Gazprom Neft became the sole owner of Sibir Energy (Mazneva 2011).

In February 2010, Gazprom Neft bought STS-Service, an oil production subsidiary of the Swedish Malka Oil, which operated in the Tomsk region. In January 2011, Gazprom Neft and TNK-BP bought the Messoyakha project from Slavneft and established a 50:50 joint venture – Messoyakha Neftegaz – to develop the Messoyakha group of fields in northern Yamal, with Gazprom Neft as the project operator (Gavshina 2011). In 2011, Gazprom Neft created a new cluster of liquid hydrocarbons production under Gazprom Neft-Orenburg, which was developing oil reserves in the eastern part of the Orenburg oil, gas and condensate field. It also received licences for the Tsarichanskaya and Kapitonovskaya group of fields located close to the Orenburg field. As a result, between 2011 and 2017, Gazprom Neft's production in the Orenburg region grew from 0.4 to 4.8 million tonnes of oil equivalent (Gazprom Neft 2018o).

During the same period, in addition to the acquisitions, Gazprom Neft brought on stream several important assets. For example, in 2012, the first phase of the Samburgskoye oil, gas and condensate field was commissioned. This field belongs to the Russian-Italian SeverEnergiya in which Gazprom Neft holds a 25% stake. In December 2013, Gazprom Neft – as the operator of the Prirazlomnoye field in the Pechora Sea – produced Russia's first Arctic offshore oil. To date, it continues to be the only Russian oil company that actually produces hydrocarbons offshore in the Russian Arctic. In 2015, the millionth tonne of Arctic Oil (known as ARCO) was extracted from the Prirazlomnoye field. The development of Prirazlomnoye is connected with much publicized Greenpeace protests in 2012 and 2013. Activists boarded the Prirazlomnaya platform on both occasions, protesting the extraction of oil in Arctic waters, claiming it to be environmentally unsafe and economically unfeasible (Greenpeace 2013).

In 2010, Gazprom transferred the licence for the Novoportovskoye oil, gas and condensate field, the largest in Yamal, to Gazprom Neft, and its subsidiary Gazprom Neft-Yamal became the operator of the Novy Port project. In 2014, development drilling started in the field. Gazprom Neft produced 2.5 million tonnes of crude in 2016 from Novoportovskoye and 5.95 million tonnes in 2017. The field is expected to yield 8 million tonnes of oil equivalent at its peak (Gazprom Neft 2019). Analysts warned that the Novoportovskoye field might be one of the first victims of Western sanctions, but Gazprom Neft managed to commission it nonetheless (Neftegazovaya vertikal 2015). The field is located about 700 kilometres from the existing (onshore) pipelines, and Gazprom Neft decided to transport Novy Port oil via the Polar Seas instead, delivering it through the Vorota Arktiki (Gates of the Arctic) offshore oil terminal (Fadeeva 2016b; Gazprom Neft 2018a). According to Gazprom Neft, zero discharge technology is used at the Vorota Arktiki terminal, which is fully automated and managed from the shore (Gazprom Neft 2016a).

In 2017, Gazprom Neft produced 62.3 million tonnes of oil, 4.3% more than 2016, thus becoming the third-largest oil producer in Russia. This growth was largely attributed to its new Arctic projects (Novy Port, East Messoyakha and Prirazlomnoye).

Like Sibneft in the 1990s, Gazprom Neft remained a trailblazer in the Russian oil sector. For example, the company requested the Ministry of Energy to include Novy Port on the list of projects subject to a new experimental tax on additional income that the Ministry had proposed to replace export duties and partially replace the mineral extraction tax. If successful, Gazprom Neft would become the first Russian oil company to transfer a major new project to this experimental tax regime (Kozlov 2018a).

Gazprom Neft also tried, albeit with limited success, to establish joint ventures with Western companies. In addition to its long-term cooperation with Shell on the Salym Petroleum Development, it formed a 50:50 joint venture with Repsol in July 2017 to develop assets in the Kondinskiy region of Khanty-Mansi Autonomous District (Gazprom Neft 2017d).

Western sanctions have forced Gazprom Neft to look for partners elsewhere, namely in the Middle East. Thus, in December 2017, Gazprom Neft signed an agreement to sell 49% of Gazprom Neft-Vostok (East) for USD 325 million to the United Arab Emirates' Mubadala Investment Company and the Russian direct investment fund RFPI. Gazprom Neft-Vostok is developing a group of mature fields in the Tomsk and Omsk regions (Kozlov 2018b).

Like Sibneft, Gazprom Neft remained investors' favourite as the fastest-growing oil company in Russia with an expansive portfolio of new projects, which enjoyed tax benefits and had a strong management team. The only disadvantage was its minimal free float, the small proportion of its shares traded in the open market (Fadeeva 2016a).

COMPANY PROFILE

Production Strategy

Strategically, Gazprom Neft aims to adapt swiftly to external challenges. In the upstream segment, it focuses on cost control, optimization of the development of mature fields, import substitution, development of new technologies to raise oil recovery ratio, commissioning of hard-to-recover reserves, implementation of major onshore (Novy Port and Messoyakha) and offshore projects (Prirazlomnoye) and improving its competencies for working with non-conventional resources (Gazprom Neft 2018b).

In 2013, the company's board of directors approved a new 2025 strategy. The document extends the previous strategy, taking into account the changing conditions in the sector and the world economy. The key strategic objective of Gazprom Neft up to 2025 is to increase shareholder value through the efficient use of its resource potential and fuller commissioning of oil reserves, developing new sources of growth and maximizing the return on investments in new projects. It expects to achieve hydrocarbon production levels of 100 million tonnes of oil equivalent by 2020 as well as refining 40 million tonnes per annum of oil in Russia (Gazprom Neft 2017c). In late 2018, the board of directors of Gazprom Neft approved a new 2030 strategy that took into account changes that occurred in 2014, mainly lower oil prices and sanctions, as well as stricter global environmental standards and technological progress. The new strategy envisages that Gazprom Neft will remain among the world's top ten publicly traded oil companies in terms of oil production (Interfax 2019).

Gazprom Neft's production strategy has two main components. The first component is to enhance its technological potential in partnership with companies such as Schlumberger, Halliburton and Shell. Under the strategy, Gazprom Neft aims to increase its oil recovery ratio from 30–35% to 55% by 2025 (Kozlov 2016). Gazprom Neft has already been successful in demonstrating quite impressive results in applying state-of-the-art technology to extract conventional reserves. It is one of the few companies in Russia that draw on chemical methods for enhancing oil recovery in mature fields (Gazprom Neft 2018b). In 2017, Gazprom Neft performed 30-stage hydrofracking in the Yuzhno-Priobskoye field in Khanty-Mansi Autonomous District, the first ever such operation in Russia (Gazprom Neft 2018b). In May 2018, Gazprom Neft signed an agreement with petroleum consultants DeGolyer & MacNaughton (with whom it had been cooperating for ten years) on the choice and application of innovative technologies for enhancing the oil recovery ratio of the Southern licensing territory of the Priobskoye field. It is expected that the oil recovery ratio there could increase by 5–6% (Gazprom Neft 2018c).

The second component of Gazprom Neft's production strategy involves commissioning new fields. This includes Prirazlomnoye on the Arctic continental shelf, the onshore Novy Port and Messoyakha projects. SeverEnergiya, a 50:50 joint venture between Gazprom Neft and Novatek, is another strategic priority for Gazprom Neft. Initially, it was an international venture involving ENI and Enel; however, following a series of share swaps in 2014, only Gazprom Neft and Novatek remained its shareholders. Arcticgaz is wholly owned by SeverEnergiya, which holds licences for the Samburgskiy block of fields. In total, recoverable reserves of SeverEnergiya amount to some 14.5 billion barrels of oil equivalent.

In 2017, Gazprom Neft was the first Russian company to cut down on its production following the agreement between Russia and OPEC. Gazprom Neft reduced its production by 2% in January 2017 compared to the 1% reduction rate achieved by most other Russian oil companies. However, the fall in Gazprom Neft's production may also be attributed to the severe cold weather that affected the output of the Novy Port and Prirazlomnoye projects and also slowed down the drilling at the company's more mature assets (Kozlov 2017a).

One of Gazprom Neft's main strategic priorities is to buy more of its supplies from Russian providers: it applies the existing Russian solutions and promotes the creation of new products in Russia. Gazprom Neft's enterprises have already been using domestic power-generating installations, onshore drilling rigs, pipes, catalysts for refining and so on (Gazprom Neft 2018d).

CSR

Sibneft was a unique case of Russian-style corporate (or rather personal) social responsibility. Roman Abramovich became Governor of Chukotka in December 2000. At the time, the region was in a sorry state. Over the previous decade, its population had decreased by a factor of 3 to 50 000 people; the outflow of residents from Chukotka was hindered only by a lack of funds to pay for flights. The indigenous population survived thanks to their reindeer; however, the headcount of reindeer fell from 500 000 in the early 1990s to 100 000 in 1999 (Proskurnina 2005).

Governor Abramovich's plan for the region had two main components. First, he radically improved the financial situation by registering the trading arms of Sibneft and Rusal in Chukotka. The profit tax from Slavneft-Trading and Sibneft-Chukotka accounted for about 60% of the regional budget revenues during Abramovich's tenure as governor. He also paid his own income tax of about USD 30 million per annum to the regional budget (Nikolaeva 2005). As a result, the total budget revenues of Chukotka increased more than eight-fold, and the tax revenue grew 40-fold. Sibneft also spent money on Chukotka via two charitable organizations: Territory and Pole of Hope (*Polyus Nadezhdy*). Besides the dramatic increase in the flows of money into the region, Abramovich also changed the budget management system by introducing the same model that has been applied in Sibneft.

As a result of these measures, Chukotka's economy began to improve. For instance, the average monthly salary rose by 3.5 times between 2000 and 2004. According to Konstantin Pulikovskiy, the authorised representative of the President in the Far Eastern District, Abramovich created the best system of regional management in Russia. Abramovich's first term in office ended in 2005, and although he was not particularly eager to remain in this position, he was elected for a second term. However, the new owner of Sibneft, Gazprom, had no intentions of helping Chukotka. A Gazprom representative once stated that 'the huge social allocations for the needs of Chukotka are the project of the former shareholders of Sibneft, not ours' (Proskurnina 2005).

Gazprom Neft performs CSR functions in a similar way to other Russian oil companies, with a focus on supporting social infrastructure and social projects in the regions where it operates. Like the other companies, it produces an annual sustainable development report, following the basic GRI standard (Gazprom Neft 2017j). The report presents Gazprom Neft's strategic management priorities in line with the UN's Sustainable Development Goals. The report outlines the company's sustainable development strategy, sustainability management and stakeholder engagement approach, corporate governance, human resource development, HSE protection, energy efficiency, supply chain responsibility, human rights, ethics and anti-corruption and community development activities. In his introductory message to the 2017 report, Alexander Dyukov made the following statement:

> Today, the company's success at the international level is measured not only by its production and financial performance but also by the way it treats its employees, the territories where it operates and the environment. For Gazprom Neft, investment in human capital, concern for the safety of all production processes, protecting nature and developing social programmes are just as important tasks as the effective implementation of business projects. (Gazprom Neft 2017j, p. 7)

Dyukov also highlighted the Biosphera treatment facilities launched at the Moscow oil refinery in 2017. Costing RUB 9 billion, the facilities purify 99.9% of the refinery's wastewater. A similar project is also being rolled out at the Omsk oil refinery (Gazprom Neft 2017j, p. 7). To honour 2017 as the Year of Ecology in Russia, Gazprom Neft supported 1100 environmental protection measures, ranging from associated gas utilization in its operations to Arctic biodiversity protection. In 2017, it invested RUB 27 billion in environmental protection measures, almost double the amount invested in 2016 (Gazprom Neft 2017j, p. 96). In 2018, investments in environmental protection amounted to RUB 19 billion (Gazprom Neft 2018q, p. 102).

In the *Sustainability Report 2018*, Dyukov made the following statement:

> In its activities, Gazprom Neft is consistently based on the principles of sustainable development. We measure success of the company not only by financial and operating indicators. Our key priorities are care about environment and prudent attitude to natural resources, safety, high technological level and systemic improvement of living standards in the regions where the company operates. (Gazprom Neft 2018q)

In the case of Messoyakha, Gazprom Neft demonstrated its ability to achieve compromises. The East Messoyakha licensing plot partially overlaps with the territory of the Messoyakha nature reserve, established to protect endangered animals and birds. Initially, there were plans to reduce the size of the reserve by some 40 000 hectares. Surprisingly, the environmental protection bodies did not particularly object to this decision because they thought that this particular area of the reserve was the least valuable in terms of flora or fauna. But the residents of the Tazovskiy region were against the proposals and protested vociferously. The shareholders of Messoyakhaneftegaz (that is, Gazprom Neft and Rosneft) responded to public opinion in September 2014 by announcing that to protect the environment, they would abandon their plans to drill wells along the Messoyakha River and they would create a no-drill buffer zone. The local population supported this initiative in public hearings (SeverPress 2014).

In 2017, Gazprom Neft formalized its policy on interaction with indigenous minorities as well as its comprehensive guidelines on interaction with indigenous peoples (Gazprom Neft 2017j, p. 124). When field development plans are being drafted for regions where indigenous peoples live, Gazprom Neft, in collaboration with the local authorities, usually organizes public hearings involving the indigenous peoples. Besides compensation measures for damage inflicted by its industrial activities, Gazprom Neft signs benefit-sharing agreements with the heads of indigenous families and provides material aid in accordance with these agreements. Over 200 such agreements were signed in 2016 (Gazprom Neft 2016b).

In 2017, the projects implemented under the socio-economic agreements (RUB 3.34 billion across five regions) included the construction of apartment buildings, a school and two sports complexes in Yamal-Nenets Autonomous District, a residential apartment block and an indoor ice-hockey rink in Khanty-Mansi Autonomous District. The grant programme (totalling RUB 25.7 million in 2017) supported projects such as an engineering skills tournament for young people, a creative skills development programme, a city festival in Khanty Mansiisk and a street art festival held in several cities (Gazprom Neft 2017j, pp. 122–34). Part of the social investment programme also focuses on building local supplier capacities to serve the industry. In 2017, Gazprom Neft concluded agreements with seven regions concerning the import substitution of lubricants and process fluids (Gazprom Neft 2017j, p. 124).

Gazprom Neft has a social investment programme called 'Native Towns'. All the company's main subsidiaries are involved in this programme. Lists of projects are compiled annually with due account for the needs of regional development and the opinion of stakeholders. The programme is implemented through cooperation agreements with the regional and local authorities and via Gazprom Neft's own social projects, a grant fund for local organizations and individual citizens, voluntary corporate actions and targeted charitable activities. In 2017, the Native Towns programme covered 35 regions with more than 2000 projects, 100 partner organizations and a total expenditure of more than RUB 20 billion (Gazprom Neft 2017j, p. 7).

A key principle of the Native Towns programme is the active involvement of Gazprom Neft's personnel in the implementation of social projects. To this end, the company also has a corporate volunteering programme which helps arrange sporting, educational and entertainment events for orphans, and children and adults with limited mobility (Gazprom Neft 2016b).

In 2018, RUB 6.8 billion were contributed by Gazprom Neft to the implementation of social projects under the Native Towns programme. By then, over 2350 social projects had already been completed, with 4773 employees acting as volunteers. The programme also has projects abroad: thus, Kustendorf CLASSIC helps young musicians and develops cultural ties between Russia and Serbia. It organizes concerts in the mountains of Serbia, and the programme consists of contests for young musicians, master-classes and performances by international classical music stars (Gazprom Neft 2018q).

In 2016, Gazprom Neft also provided financial support to the Russian Geographical Society, the Hermitage, the Saint Petersburg Yacht Club, the Russian Military and Historic Society, the Yamal Cooperation Foundation, UNESCO's International Assistance Foundation in Moscow, the Saint Petersburg Union of Journalists, the Federation of Figure-Skating in Yamal-Nenets Autonomous District and so on (Gazprom Neft 2016b).

Offshore

Gazprom Neft is the only Russian company that produces Arctic offshore oil, having commissioned the Prirazlomnoye field in the Pechora Sea in December 2013. However, its other Arctic offshore licences and its licences in the Sea of Okhotsk off Sakhalin Island are still only at the exploration stage.

Gazprom Neft recognizes that it needs foreign partners for the development of these fields; however, because of the Western sanctions in connection with the conflict in Ukraine, it mainly seeks potential allies in Asia. Thus, in 2016, Gazprom Neft and China National Offshore Oil Corporation (CNOOC) discussed opportunities for the joint development of offshore fields in Russia. In March 2017, during the Arctic Forum in Archangelsk, Gazprom Neft and Indian ONGC signed a framework agreement on Arctic offshore cooperation, focusing on exploration opportunities in the Dolginskoye field in the Pechora Sea. However, there are doubts about the capability of Chinese or Indian companies to replace the international oil companies in the challenging Arctic offshore ventures.

In 2017, Gazprom Neft-Sakhalin drilled the first well at the Ayashskiy block off Sakhalin Island in the Sea of Okhotsk. Originally Gazprom held the licence for this plot; however, in 2017, it transferred it to Gazprom Neft, which had managed to convince the parent company of the high oil-bearing potential of the plot (Kozlov 2017b). In October 2017, Gazprom Neft-Sakhalin discovered the Neptun field in the Sea of Okhotsk with geological reserves estimated at 255 million tonnes of oil equivalent (Alekseev 2017), and in November 2018, it discovered another field, Triton, with geological reserves estimated at 137 million tonnes of oil equivalent (Gazprom Neft 2018p). The company subsequently began seeking a partner to develop the Neptun field, in which Shell reportedly expressed interest (Neftianka 2018). The sea at the Ayashskiy block is only 62 metres deep; therefore, the project cannot be formally subject to the US sanctions that cover deep-water projects. Despite this, foreign companies proceed with caution about getting involved in such ventures. Following the discovery of Neptun, Alexander Dyukov confirmed that 'the development of the offshore assets of Gazprom Neft remains a strategic priority for the company' (Gazprom Neft 2017e).

Gazprom Neft-Sakhalin holds licences for four Russian Arctic offshore plots: Severo Vrangelevskiy (Eastern Siberian and Chukchi Seas), Kheisovskiy (Barents Sea) as well as Dolginskiy and Severo-Zapadniy plots (Pechora Sea). In 2015, Gazprom Neft lobbied for changes to the Dolginskoye field licence because it was not satisfied with the results of exploration drilling. Production in this field is now due to begin in 2031 rather than 2019 as was originally planned (Podobedova 2015).

Shale Oil

Gazprom Neft has expended significant amounts of capital on the development of hard-to-recover oil reserves. In total, the company has some 527 million tonnes of hard-to-recover reserves mostly in areas developed by its subsidiaries Gazprom Neft-Khantos in Khanty-Mansi Autonomous District and Gazprom Neft-Muravlenko in Yamal-Nenets Autonomous District (Gazprom Neft 2017i).

Experts from Gazprom Neft believe that the Bazhenov shale formation looks more attractive than the Arctic continental shelf east of the Urals mountains. Bazhenov is the world's largest shale oil formation. It is a 20–30 metre thick layer of rock at a depth of over two kilometres, extending throughout much of Western Siberia – a region that already has well-developed petroleum infrastructure. According to some estimates, its resources exceed 140 billion tonnes; however, it is characterized by a low oil recovery ratio of only 6–7% (Korostikov and Tarasenko 2019).

In September 2017, Gazprom Neft and the government of Khanty-Mansi Autonomous District signed an agreement to establish the Bazhenov Technological Centre in Khanty-Mansi Autonomous District. The parties agreed to cooperate on this national project to create domestic technology and equipment to develop the Bazhenov formation. The Bazhenov Centre will concentrate Gazprom Neft's state-of-the-art competencies and technologies as a basis for developing economically efficient methods and technologies. An area of the Krasnoleninskoye field has been designated to be used as a testing site for new technologies and equipment. In May 2017, a Gazprom Neft project to study the Bazhenov formation received national status with support from the Ministry of Energy and the Ministry of Natural Resources (Gazprom Neft 2017f).

Continuing this trend, in May 2018, Gazprom Neft, the Ministry of Industry and Trade and the government of Khanty-Mansi Autonomous District signed a MoU concerning the development of Bazhenov. Gazprom Neft will make investments and coordinate the activities of the technological partners. While the Ministry will encourage the signing of special investment contracts with technological partners, the government of Khanty-Mansi Autonomous District will create a favourable fiscal regime for these contracts. Gazprom Neft expects to produce 2.5 million tonnes of oil per annum from Bazhenov by 2025 (Gazprom Neft 2018h).

In October 2017, Gazprom Neft and the Engineering Centre of the Moscow Physical and Technical Institute established a Centre of R&D Support for Hydrofracking in Saint Petersburg to develop optimal technological solutions for Bazhenov (Gazprom Neft 2017g). Gazprom Neft-Orenburg has also begun studying Domanicoid sediments, which are reserves of unconventional hydrocarbons in tight shale layers with forecast resources in the region of some 1 billion tonnes of oil (Gazprom Neft 2018o).

Innovation

Gazprom Neft's Innovative Development Programme aims to create technologies to enhance well productivity, develop the Bazhenov formation, improve oil recovery in depleted fields and manufacture catalysts for refining; the programme also aims to digitalize oil production, refining and marketing (Gazprom Neft, 2017j, p. 23).

A large part of Gazprom Neft's investment in innovation is focused on digital transformation (Gazprom Neft 2017j, p. 25). In April 2018, Gazprom Neft established a directorate for digital transformation led by Chief Digital Officer (CDO) Andrei Belevtsev. The directorate will work on creating a unified system of digital projects to radically improve the operating efficiency of all business processes and develop in-house services. Gazprom Neft plans to create an in-house IT platform to support services for generating new data flows and volumes and developing instruments for predictive analysis. The company expects to begin developing new products and services based on the introduction of state-of-the-art digital technologies.

In 2017–18, the company implemented a series of projects involving blockchain technology, artificial intelligence, predictive analysis, big data and the so-called digital twins of oilfields (Gazprom Neft 2018k). 'Digital twins' are virtual models of oilfields that change in tandem with regularly updated information which is passed from site-based sensors. The digital twins enhance the ability to predict technical issues and make timely decisions about equipment repairs, thus reducing operational costs (Gazprom Neft 2017j, p. 23).

Gazprom Neft aims to create the first 'digital plant' as part of a project to establish Russia's first digital platform for managing logistics, production and sales of petroleum products. The company has been working on this since 2017 and is cooperating with leading Russian IT companies and institutions, such as

Tsifra, Skolkovo Institute of Science and Technology and Moscow Institute of Physics and Technology (Gazprom Neft 2018l, 2018m).

In winter 2017–18, Gazprom Neft also tested an innovative and environmentally friendly approach to seismic exploration, using the green seismic technology at the Zapadno-Pokurskoy area in Khanty-Mansi Autonomous District. Besides reducing the anthropogenic impact on the environment, this technology draws on a wireless system of data transmission to enable the study of areas that are inaccessible by standard methods. Another pilot project (under Green Seismic 2.0) has been implemented in the same area. This project promotes the use of light drilling rigs and snowmobiles instead of heavy machinery as a way to avoid cutting down trees during seismic testing (Gazprom Neft 2018n).

Internationalization

In its 2025 strategy, which aimed to produce 100 million tonnes of oil equivalent per annum, Gazprom Neft predicted that its foreign upstream operations will account for at least 10% and places particular emphasis on Iraq and Venezuela (Neftegaz 2013). However, since investment risk in both countries is very high, due to political instability, Gazprom Neft has been searching for other attractive opportunities, albeit with somewhat mixed results.

One of Gazprom Neft's most important international upstream acquisitions was made in December 2008 when it purchased a 51% share of the Serbian company Naftna Industrija Srbije (NIS) for EUR 400 million when NIS was privatized. The government wholly owned NIS with fields in Serbia and Angola as well as two refineries and 486 fuel stations in Serbia (Tutushkin 2008b). In 2011, Gazprom Neft bought another 5.15% of NIS, raising its stake to 56.15%. Serbia is the only European country in which Gazprom Neft produces oil. Via NIS, Gazprom Neft also operates in Hungary and Romania, albeit at the exploration stage.

From 2008, hoping to make its presence felt in the Middle East, Gazprom Neft initially looked to Iran. In November 2009, it signed a MoU with the National Iranian Oil Company (NIOC) concerning the Anaran block in western Iran (Tutushkin 2009a). However, being dissatisfied with Gazprom Neft's passivity, NIOC terminated the MoU in 2011 and started seeking new partners. Tehran allegedly had the impression that since Gazprom Neft produced oil in Iraq, Iran was of little genuine interest to it (Raibman 2011). However, according to other reports, Gazprom Neft dragged its feet and eventually gave up the project because Gazprom proper was not keen to work in a country under international sanctions (Melnikov and Gabuev 2011).

A major upstream breakthrough for Gazprom Neft was achieved in Iraq. In January 2010, the Iraqi government signed a contract with the winner of an international tender to develop the Badra project in eastern Iraq. The consortium comprised Gazprom Neft (30%, operator), Kogas (22.5%), Petronas (15%) and Turkish Petroleum (TPAO) (7.5%), with the Iraqi Oil Exploration Company (OEC-Iraq) keeping a 25% share. In 2014, Gazprom Neft launched the commercial development of the Badra oilfield. This was the first major international E&P project that it started from scratch. In 2012, Gazprom Neft secured further projects in Iraq (Gazprom Neft 2018f). In 2013, Gazprom Neft expanded its presence to Kurdistan in northern Iraq where it became the operator of a project (with an 80% stake) to develop the Halabja block with estimated reserves of some 100 million tonnes of oil (Solodovnikova 2013a).

Venezuela has been another potential growth area for Gazprom Neft. In 2008, the National Oil Consortium was established in which Gazprom Neft was an equal partner with Rosneft, LUKOIL, TNK-BP and Surgutneftegas. In spring 2010, the joint venture PetroMiranda was set up to develop the Junin-6 block in Venezuela's Orinoco River basin. PetroMiranda comprised the National Oil Consortium with 40% and Corporacion Venezolana del Petroleo, a subsidiary of Venezuela's PDVSA, with 60%. Forecasts of recoverable reserves of Junin-6 are estimated at 10.96 billion barrels of oil, and the first oil was produced in 2012 (Gazprom Neft 2018g). Initially, Gazprom Neft was appointed to coordinate the activities of all the Russian companies; however, later Rosneft took over this role. Gradually, other companies quit the National Oil Consortium, leaving behind Rosneft (80%) and Gazprom Neft (20%).

Gazprom Neft sought to introduce itself as a strong player in international offshore projects, but without much success. In 2010, it concluded an agreement with Petronas and began working on the Cuban continental shelf. In July 2011, it signed a contract to cooperate with Petronas and the state-owned Cuba Oil Union (CUPET). However, having spent some USD 12 million, Gazprom Neft decided to abandon the project on the Cuban continental shelf in February 2013 without having found any oil (Solodovnikova 2013b).

In mid-2010, Gazprom Neft, Guinean GEPetrol and the Ministry of Energy of Equatorial Guinea signed a production-sharing agreement relating to two oil blocks on the continental shelf of Equatorial Guinea. It was to be the first Russian oil project in Equatorial Guinea, where US companies had hitherto dominated (Peretolchina et al. 2010). However, in 2014, Gazprom Neft left the project, deeming it insufficiently attractive (Mordyushenko 2014).

Gazprom Neft also tried to secure a stake in the (onshore) Elephant Field in south-west Libya. Gazprom reached a preliminary agreement with ENI in 2008 during a multi-stage asset swap with the Italian company; however, because of the hostilities in Libya, Gazprom Neft's expectations did not materialize.

Downstream, Gazprom Neft focuses on expanding its niche in the fuel markets of the CIS. In 2006, Gazprom Neft entered the Central Asian retail market and founded a subsidiary, Gazprom Neft-Asia. In 2006, Gazprom Neft bought some 100 fuel stations in Kyrgyzstan for USD 99 million. It became the second Russian oil company (after Alliance) to enter the Kyrgyz market (Tutushkin 2006). In 2010, Gazprom Neft bought a network of 20 fuel stations from the Kazakh company ARNA Petroleum along with nine land plots for the construction of new fuel stations (Tutushkin and Vasilev 2010).

Gazprom Neft's advance downstream in Europe was not particularly successful after its initial breakthrough with NIS in Serbia. Gazprom Neft was interested in Greece and hoped to participate in the privatization of a stake in Hellenic Petroleum in 2012 (Melnikov et al. 2012). It also wished to gain a foothold in Albania and planned to take part in the privatization of Albpetrol (Mazneva 2012). However, neither of these ambitions materialized. Gazprom Neft, however, was more fortunate in Italy. In early 2009, Gazprom Neft bought the oil- and lubricant-producing plant Chevron Italia, located in the port city of Bari on the Adriatic Sea, with a capacity of 30 000 tonnes of oil and 6000 tonnes of lubricant per annum (Tutushkin 2009b).

Gazprom Neft also tried to gain a foothold downstream in Asia. In November 2013, during Vladimir Putin's visit to Vietnam, Gazprom Neft signed a framework agreement to purchase part of the Dung Quat refinery from Petrovietnam (Serov and Chelpanova 2013). However, in January 2016, it broke off negotiations with Petrovietnam because it was dissatisfied with the position of the Vietnamese Ministry of Industry and Trade, which did not intend to retain preferential import tariffs for the refinery (Neftegazovaya vertikal 2016).

COPING WITH THE CHALLENGE OF CHANGE

Crises, Oil Prices and Sanctions

Like other Russian companies, Sibneft/Gazprom Neft is seriously affected every time the oil prices fall. When the oil prices dropped in 1998, and the Russian financial system crashed, Sibneft engaged in cost reduction. It cut the investment programme of Noyabrskneftegaz by drastically reducing its considerable expenditures, which included capital construction and drilling. It also decided to uncouple Noyabrskneftegaz from its oil service facilities (Neft, gaz i fondovyy rynok 2008).

In the aftermath of the 2008 oil price collapse, Gazprom Neft lost more than USD 20 billion of its year-end capitalization. However, Gazprom Neft's response was somewhat interesting: It bought out 0.44% of its shares to support liquidity, obtaining them at minimum prices. The company expected that once the crisis had passed, the shares could be sold or used for the options programme for managers (Mazneva 2009). Despite the crisis, Gazprom Neft doubled its expenditure on the 1286 staff at its headquarters in 2010 (Gavshina

2010). It is not easy to determine whether this development was driven by a proactive human resources policy or reflected a principal-agent problem.

Gazprom Neft is of course well aware of the importance of oil price dynamics and identifies the following macroeconomic factors that affect its work: changes of market prices of oil and petroleum products, dynamics of the rouble/dollar exchange rate and inflation, taxation and changes in tariffs for the transportation of oil and petroleum products (Gazprom Neft 2018b).

The company's leadership closely monitors the oil price landscape and maps out its strategy accordingly. Hence, in 2016, Alexander Dyukov announced that the company assumed that the oil prices would be USD 50–60 per barrel over three years. The initial forecast of the company for 2016 was USD 25–30 per barrel. Dyukov emphasized that the company could comfortably operate at USD 25 per barrel (Fadeeva 2016).

In its *2016 Annual Report*, Gazprom Neft stated that the dynamics of supply and demand, the reduction of excess oil in the market and changes in the tactics of oil producers mitigated the risks of price volatility in the short term. The company expected that the following factors would impact the oil market in 2017: a rise in the world's oil consumption, the OPEC+ agreement, the dynamics of oil production in the United States and the geopolitical limitations of oil production in certain OPEC member countries (Gazprom Neft 2018b).

The attitude of the company towards the challenges of oil price volatility and sanctions is well described by Sergei Vakulenko, Head of Gazprom Neft's Department of Strategy and Innovations. According to him, before 2014 (that is, in the pre-sanctions period) raising capital was easy, and the whole oil sector was in a hurry to accumulate as many assets as possible, believing that whoever expanded fastest would win (Vakulenko 2018). In this respect, claims Vakulenko, Gazprom Neft was fairly unique. It demonstrated the highest growth rates in the sector, and they were achieved because of a well-thought-out plan. Gazprom Neft calculated from the very beginning how many and what projects it needed at a specific stage of its development; when they were accomplished, only then were new projects chosen. This principle guided the company before 2014, and it remains relevant today. However, Gazprom Neft undoubtedly also had to revise some of its projects given the altered economic conditions. For example, the upgrading of refineries was slightly delayed due to budget constraints. The company changed its approach to high-cost production projects; it did not give them up completely but tried to cut costs as much as possible. In general, the business model that Gazprom Neft adopted turned out to be viable and survived the acid test of sanctions.

Vakulenko commented that due to the specifics of the domestic revenue-based tax system, Russian oil companies learned how to work efficiently with very low oil prices. Even when the oil prices were above USD 100 per barrel, for the Russian oil companies, the prices did not exceed USD 35–40.

> If we speak about the next 2–3 years, such specific things as wars, revolts, natural disasters, that is any global disturbances in oil-producing regions, might significantly affect the production potential. Recall that the period of the great growth of oil prices that resulted in USD 146 per barrel in 2008 began with Hurricane Katrina in 2003. From a different perspective, that period was also characterized by high economic growth that pushed oil demand upwards. Now, an unstable global political situation contributes to the dampening of economic activity. (Gazprom Neft 2018e)

Sanctions and low oil prices threatened Gazprom Neft's most ambitious upstream projects, such as Prirazlomnoye. However, in late 2014, Gazprom Neft was confident that this offshore project would be profitable even with a considerable decline in oil prices. Interestingly, in early 2014, Alexander Dyukov commented that the fiscal benefits granted to the project would ensure the efficiency of the Prirazlomnoye development even if oil prices dropped to USD 80 per barrel (Neftegazovaya vertikal 2015). The subsequent events, however, showed that he was overly optimistic in his oil price forecast (albeit realistic in assessing the value of state support).

To survive low oil prices, Gazprom Neft successfully lobbied for significant fiscal benefits: the mineral production tax was reduced to zero for the Prirazlomnoye field; this waiver was initially provided up to 2019 but was later extended to 2022. From April 2014 a lower rate of export duty was also granted for Prirazlomnoye oil. In July 2014, the Ministry of Natural Resources proposed transferring Prirazlomnoye from the second to the third category of project complexity, which meant the mineral production tax would be reduced from 15% to 10% after a certain level of production was achieved. Also, a tax manoeuvre helped the project by lowering even more export duty, thus making it possible for Prirazlomnoye to succeed in the face of low oil prices (Neftegazovaya vertikal 2015).

Still, in March 2016, Gazprom Neft wrote a letter to the Ministry of Natural Resources complaining about the adverse effects of the sanctions and declining oil prices on the company's ability to invest in offshore exploration and demanding to have financial benefits for its offshore exploration projects (Starinskaya 2016).

In 2017, a new budget rule for the Russian petroleum sector was introduced. In 2018, all additional income generated when the oil prices exceeded USD 40 per barrel was to be channelled to a special reserve fund. Gazprom Neft adhered to this rule. When approving the business plan for the subsequent three years, Gazprom Neft proceeded from the forecast macroeconomic indicators determined by the Ministry of Economic Development: an expected price of USD 44 per barrel for Urals oil and a USD/RUB exchange rate of 64. All additional income was channelled to the fund designated by the budget rule; the company would use it to repay creditors, pay dividends and finance its investment programmes. Experts believed that these accumulations might be necessary for Gazprom Neft to reduce risks, presumably because of the sanctions (Petlevoy et al. 2018).

In early February 2018, Alexander Dyukov said that the scenario whereby the participants of the OPEC+ agreement would make greater oil production cuts seemed unacceptable. 'We would still hope that the production quotas would increase. You see that demand is growing, the price is also growing. The market is close to stabilization; therefore, if we proceed from what we observe now, I will accept this scenario' (Dyukov, cited in Starinskaya 2018).

When Western countries imposed sanctions on Russia and its petroleum sector over the conflict in Ukraine, Gazprom Neft began searching for partners for its offshore projects in the East. 'After the introduction of the sanctions regime, these activities [in offshore fields] became more complicated, but the Eastern direction – the direction of the Asia-Pacific region – permits saying that the companies are interested or express interest in our projects, and we hope to work in this direction', said Andrei Patrushev, the Deputy Director of Gazprom Neft for the development of offshore projects, in summer 2016 (Petlevoy 2017).

Western sanctions also had an unexpected indirect impact on Gazprom Neft: taking advantage of the growing anti-Russian sentiment in Europe, Serbia tried to revise the terms for the privatization of NIS. The Serbian Ministry of Internal Affairs began to investigate the privatization deal (Neftegazovaya vertikal 2015). However, because of Gazprom Neft's political connections and support from the Russian government, it managed to weather the storm.

Climate Change

Gazprom Neft is working on the assumption that hydrocarbons will continue to play a key role in the world's economy in the future, supplying some 90% of the primary energy. In assessing the global dynamics of hydrocarbon supply and demand, the company uses data from the US Energy Information Administration and the IEA in its *2017 Annual Report*.

Like many of its Russian counterparts, Gazprom Neft does not seem to consider climate change a serious issue. Its *2016 Sustainability Report* mentions the word 'climate' only twice, once on page 36 in connection with the UN's

Sustainable Development Goals (Gazprom Neft 2016b) and another time in a very general context:

> Gazprom Neft is one of the leaders of the Russian oil sector and recognizes its responsibility for preserving the natural environment for the current and future generations. The problem of climate change strengthens the importance of this issue and activities aimed at reducing emissions at the level of the company and the country as a whole. Solving this task, the company implements the programme to raise the level of APG utilization at all its oil-producing assets.

Gazprom Neft's attitude towards renewable energy is reflected in an interview with Sergei Vakulenko, Head of the Department of Strategy and Innovations:

> If we speak about alternative energy: solar, air and water, electric vehicles and so on, we certainly take them into account in our strategy but in a much longer-term perspective. The speed of this development will be determined by a whole range of factors, such as the rates of improvement of battery-manufacturing technologies, the desire of governments to carry out social engineering, prompting people to make a transition to electric cars and the readiness of people to do this. In a very extreme scenario, people from developed countries may decide that global warming is exceptionally serious and to mitigate it, they will have to sacrifice part of their income so that growing car ownership in the developing countries will follow the electric path. Under such developments, demand for oil would indeed decrease more rapidly than what people envisage now. But this is on the condition that the countries of the first world are rich enough to agree to this, and it involves a whole spectrum of factors, such as technological readiness and the availability of raw materials and not only lithium, about which everybody seems to know, but also cobalt, neodymium for permanent magnets and so on. (Gazprom Neft 2018e)

Vakulenko believes that peak demand for oil in 2040–50 is quite possible and probable: 'The trend will be determined by the degree to which people will be concerned about environmental issues' (Gazprom Neft 2018e).

Gazprom Neft monitors GHG emissions generated through its activities: both direct and indirect GHG emissions related to third-party consumption of electricity, heat, steam and so on (Gazprom Neft 2016b). According to its *2016 Sustainability Report*, the total volume of GHG emissions in 2016 amounted to 16 million tonnes of CO_2 equivalent, a 28% rise from 2015. This increase is due to the commissioning of major new fields where infrastructure for utilizing associated petroleum gas (APG) would be built later on; the rise in GHG emissions is also ascribable to the increased consumption of heat and electricity because of an unusually cold winter. In 2018, the company's GHG emissions amounted to 19 million tonnes of CO_2 equivalent due to increased APG flaring and oil refining (Gazprom Neft 2018q, p. 105) (Figure 4.2).

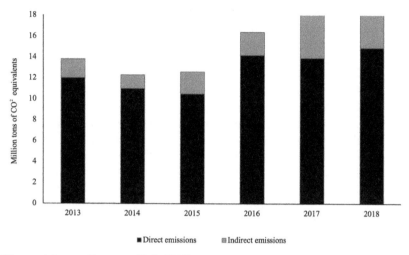

Figure 4.2   Gazprom Neft GHG emissions

Source: Gazprom Neft (2016b, p. 98, 2018q, p. 105).

In practice, the utilization of APG is Gazprom Neft's main contribution to climate change mitigation. Already in 2012, it signed an agreement with SIBUR on implementing the Noyabrsk Integrated Project for the Vynagapurovskaya group of five fields in Yamal-Nenets Autonomous District. APG produced from these fields was to be delivered to the newly built Vynagapurovskiy gas-processing plant of SIBUR (Neftegaz 2012). Cooperation with SIBUR in utilizing APG continued: in September 2015, Gazprom Neft and SIBUR commissioned the Yuzhno-Priobskiy gas-processing plant with a capacity of 900 million cubic metres of APG per annum (Gazprom Neft 2015). Continuing this trend, in April 2018, Messoyakhaneftegaz started a unique practice of injecting APG from East Messoyakha field into the gas cap of the West Messoyakha field. This is also a way of storing gas for future use (Gazprom Neft 2018i).

In February 2018, Gazprom Neft began testing the installation of a complex gas treatment system and its injection into reservoirs in the Novoportovskoye field. This installation will permit the utilization of 95% of the field's APG. Processed APG from the installation will also be used as fuel for the largest gas-turbine power station in Yamal (Gazprom Neft 2018j). Gazprom Neft seeks to enhance the energy efficiency of its operations and optimize the utilization of energy resources. This work is incorporated into the company's energy strategy and follows ISO 50001:2011 (Gazprom Neft 2016b).

Gazprom Neft also aims to increase APG utilization in its foreign operations. Thus, in December 2017, the company began commercial operation of the installation for gas processing at the Badra field. Dry gas prepared at the Badra field is transported to the Az-Zubaidiya power station; gas is also used for the needs of the Badra project as fuel for a gas-turbine power station. As Alexander Dyukov noted, the installation will enable the monetization of all hydrocarbons produced in the field and ensure a 95% level of APG utilization (Gazprom Neft 2017h).

Through its Serbian subsidiary NIS, Gazprom Neft started experimenting with the renewables business. Since 2011, NIS has been transforming itself from an oil and gas company into an energy holding which develops, among other things, renewables (geothermal and wind). NIS uses geothermal energy for heating purposes. In 2016, NIS also completed the site for the construction of its own wind farm in Plandishte with a capacity of 100 MW consisting of 40 wind generators (Gazprom Neft 2016b). Gazprom Neft also has very small-scale wind and solar projects in Yamal where some remote sites consume electricity generated by wind, solar and diesel installations combined.

CONCLUSION

Gazprom Neft is the result of the transformation of the former 'robber baron' Sibneft, which was involved in several major corporate controversies in Russia, into a subsidiary of Gazprom, which is better known for its consistency and stability as the blue whale of the Russian gas sector. Gazprom Neft continues to manifest a fairly entrepreneurial approach to the oil business, differentiating it from its parent company. It has also proved to be a doer that has managed to launch the three largest Arctic oil projects in Russia (including the first ever Russian Arctic offshore project). Having received strong support from the state, it has overcome macroeconomic problems and geopolitical challenges. It is also a digital pioneer of the Russian petroleum sector.

Although the posts of the chairman of the board of directors and the company director passed between different people, partly reflecting changes in Russia's political configuration, the leadership of Sibneft was dominated by the larger-than-life characters of Roman Abramovich and Boris Berezovskiy. Since being taken over by Gazprom, the leadership of Gazprom Neft has become more stable. Alexander Dyukov has been the CEO and chairman of the management board of Gazprom Neft since 2008.

Gazprom Neft turned its back on the patronage that Sibneft had shown to Russia's Far Eastern province of Chukotka and continues its CSR activities along the more traditional lines of Russian oil companies, with a focus on social projects in the regions of its operations. On the issue of climate change, Gazprom Neft holds a somewhat sceptical position, expecting hydrocarbons to dominate future development scenarios; despite this, it has made considerable investments in operational efficiency and utilization of APG.

NOTE

1. Use of pressurized water to maximize oil extraction.

REFERENCES

Alekseev, A. (2017), 'Yavlenie Neptuna', accessed 16 October 2018 at http://www.gazprom-neft.ru/press-center/sibneft-online/archive/2017-october/1205484/.

Baraulina, A., I. Malkova and A. Tutushkin (2009), 'Inostrantsy prodali Sibir', accessed 15 October 2018 at https://www.vedomosti.ru/newspaper/articles/2009/04/24/inostrancy-prodali-sibir.

Borisov, N. (2005), 'Sibneft budet schedroy', accessed 15 October 2018 at https://neftegaz.ru/press/view/1958.

Derbilova, E. (2005), 'Millhouse pomog Gazpromu', accessed 15 October 2018 at https://www.vedomosti.ru/newspaper/articles/2005/09/13/millhouse-pomog-gazpromu.

Egorova, T. (2006a), 'Sibneft nashla kadry', accessed 15 October 2018 at https://www.vedomosti.ru/newspaper/articles/2006/04/04/sibneft-nashla-kadry.

Egorova, T. (2006b), 'Kak Abramovichu', accessed 15 October 2018 at https://www.vedomosti.ru/newspaper/articles/2006/05/24/kak-abramovichu.

Fadeeva, A. (2016a), 'Gazprom Neft khochet ostatsya sredi liderov po neftedobyche', accessed 15 October 2018 at https://www.vedomosti.ru/business/articles/2016/06/14/645165-gazprom-neft.

Fadeeva, A. (2016b), 'Gazprom Neft otpravila pervyi tanker nefti s Novoportovskogo mestorozhdeniya', accessed 15 October 2018 at https://www.vedomosti.ru/business/articles/2016/05/26/642469-gazprom-neft-novoportovskogo.

Forbes (2011), 'Abramovich: Slukhi o Berezovskom kak aktsionere Sibnefti pomogali kompanii', accessed 15 October 2018 at http://www.Forbes.ru/news/75963-abramovich-sluhi-o-berezovskom-kak-aktsionere-sibnefti-pomogali-kompanii.

Gavshina, O. (2010), 'Neftyanye kompanii – lidery po rostu raskhodov na personal', accessed 16 October 2018 at https://www.vedomosti.ru/management/articles/2010/08/18/platyat_bolshe.

Gavshina, O. (2011), 'Sovmestnoe predpriyatie Gazprom Nefti i TNK-BP vykupit proekt u dochki', accessed 15 October 2018 at https://www.vedomosti.ru/business/articles/2011/01/12/podelili_neft.

Gazprom Neft (2016a), 'Nachalas otgruzka Yamalskoy nefti cherez morskoi terminal "Vorota Arktiki"', accessed 23 January 2019 at https://www.gazprom-neft.ru/press-center/news/1113215/.

Gazprom Neft (2016b), *Gazprom Group's Sustainability Report 2016*, accessed 16 October 2018 at http://www.gazprom-neft.com/social/reports/.

Gazprom Neft (2017a), 'Istoriya kompanii', accessed 11 October 2018 at http://www.gazprom-neft.ru/company/history/.

Gazprom Neft (2017b), 'Geologorazvedka i dobycha nefti i gaza', accessed 11 October 2018 at http://www.gazprom-neft.ru/company/business/exploration-and-production/.

Gazprom Neft (2017c), 'Gazprom Neft vkratse', accessed 11 October 2018 at http://www.gazprom-neft.ru/company/at-a-glance/.

128 *Russian oil companies in an evolving world*

128 *Russian oil companies in an evolving world*

128 *Russian oil companies in an evolving world*

128 *Russian oil companies in an evolving world*

128 *Russian oil companies in an evolving world*

128 *Russian oil companies in an evolving world*

Gazprom Neft (2017d), 'Sovet direktorov Gazprom Nefti rassmotrel voprosy strategicheskogo razvitiya kompanii', accessed 15 October 2018 at http://www.gazprom-neft.ru/press-center/news/1247213/.

Gazprom Neft (2017e), 'Gazprom Neft otkryla novoe mestorozhdenie na shelfe Okhotskogo morya', accessed 16 October 2018 at http://www.gazprom-neft.ru/press-center/news/1166741/.

Gazprom Neft (2017f), 'Gazprom Neft i pravitelstvo KHMAO podpisali soglashenie o sozdanii Technologicheskogo Centra Bazhen', accessed 16 October 2018 at http://www.gazprom-neft.ru/press-center/news/1133299/.

Gazprom Neft (2017g), 'Gazprom Neft i MFTI sozdali nauchno-technicheskiy tsentr dlya soprovozhdeniya i izucheniya GRP v Bazhenovskoy svite', accessed 16 October 2018 at http://www.gazprom-neft.ru/press-center/news/1189202/.

Gazprom Neft (2017h), 'Gazprom Neft vvela v promyshlennuyu ekspluatatsiu gazovyi zavod na mestorozhdenii Badra v Irake', accessed 16 October 2018 at http://www.gazprom-neft.ru/press-center/news/1263566/.

Gazprom Neft (2017i), 'V Gazprom Neft-Khantose sozdan tsentr upravleniya dobychey, ispolzuyuschiy technologiu tsifrovykh dvoinikov', accessed 16 October 2018 at http://www.gazprom-neft.ru/press-center/news/1226891/.

Gazprom Neft (2017j), *Aiming Higher: 2017 Sustainability Report*, accessed 9 February 2019 at http://ir.gazprom-neft.com/fileadmin/user_upload/documents/annual_reports/gpn_csr2017_eng_200718.pdf.

Gazprom Neft (2018a), 'Proekt Novyy Port', accessed 15 October 2018 at http://www.gazprom-neft.ru/company/business/exploration-and-production/new-projects/new-port/.

Gazprom Neft (2018b), *Gazprom Neft 2017 Annual Report*, accessed 15 October 2018 at http://ir.gazprom-neft.com/fileadmin/user_upload/documents/annual_reports/gpn_ar17_eng.pdf.

Gazprom Neft (2018c), 'Gazprom Neft povyshaet effectivnost razrabortki Priobskogo mestorozhdeniya vmeste s DeGolyer and McNaughton', accessed 15 October 2018 at http://www.gazprom-neft.ru/press-center/news/1646773/.

Gazprom Neft (2018d), 'Sovet direktorov Gazprom Nefti rassmotrel mery po minimizatsii doli importnykh zakupok', accessed 16 October 2018 at http://www.gazprom-neft.ru/press-center/news/1638994/.

Gazprom Neft (2018e), 'Gazprom Neft: Osnovnye idei strategicheskogo razvitiya', accessed 16 October 2018 at http://www.gazprom-neft.ru/press-center/lib/1642212/.

Gazprom Neft (2018f), 'Gazprom Neft vvela v ekspluatatsiyu vtoruyu skvazhinu na mestorozhdenii Sarqala', accessed 16 October 2018 at http://www.gazprom-neft.ru/press-center/news/1526676/.

Gazprom Neft (2018g), 'Mezhdunarodnye proekty', accessed 16 October 2018 at http://www.gazprom-neft.ru/company/business/exploration-and-production/international-projects/.

Gazprom Neft (2018h), 'Gazprom Neft, Minpromtorg i pravitelstvo KHMAO-Ugry podtverdili sotrudnichestvo v proekte osvoeniya Bazhenovskoy svity', accessed 16 October 2018 at http://www.gazprom-neft.ru/press-center/news/1621476/.

Gazprom Neft (2018i), 'Messoyakhaneftegaz realizuet unikalnyi proekt po utilizatsii poputnogo neftyanogo gaza', accessed 16 October 2018 at http://www.gazprom-neft.ru/press-center/news/1509719/.

Gazprom Neft (2018j), 'Gazprom Neft nachala puskonaladochnye raboty na ustanovke kompleksnoy podgotovki gaza Novoportovskogo mestorozhdeniya', accessed 16 October 2018 at http://www.gazprom-neft.ru/press-center/news/1442588/.

Gazprom Neft (2018k), 'Tsifrovizatsiya stanet novym biznes-napravleniem Gazprom Nefti', accessed 16 October 2018 at http://www.gazprom-neft.ru/press-center/news/1509691/.

Gazprom Neft (2018l), 'Gazprom Neft sozdaet pervyi tsifrovoy zavod', accessed 16 October 2018 at http://www.gazprom-neft.ru/press-center/news/1621474/.

Gazprom Neft (2018m), 'Gazprom Neft sozdaet sobstvennuyu tsifrovuyu technologiu razrabotki mestorozhdeniy', accessed 16 October 2018 at http://www.gazprom-neft.ru/press-center/news/1526739/.

Gazprom Neft (2018n), 'Gazprom Neft rasshiryaet primenenie zelenykh technologiy v seismorazvedke', accessed 16 October 2018 at http://www.gazprom-neft.ru/press-center/news/1595456/.

Gazprom Neft (2018o), 'Sovet Direktorov Gazprom Nefti rassmotrel vopros o perspektivakh razvitiya Orenburgskogo neftedobyvayuschego klastera', accessed 15 October 2018 at http://www.gazprom-neft.ru/press-center/news/1509639/.

Gazprom Neft (2018p), 'Gazprom Neft otkryla vtoroye mestorozhdeniye na shelfe Okhotskogo Morya', accessed 29 June 2019 at https://www.gazprom-neft.ru/press-center/news/2021843/.

Gazprom Neft (2018q), *Sustainability Report 2018*, accessed 3 December 2019 at https://csr2018.gazprom-neft.ru/.

Gazprom Neft (2018r), 'Gazprom Neft at a glance', accessed 3 December 2019 at https://www.gazprom-neft.ru/company/about/at-a-glance/.

Gazprom Neft (2019), 'Proetk Novy Port', accessed 23 January 2019 at https://www.gazprom-neft.ru/company/business/exploration-and-production/new-projects/new-port/.

Greenpeace (2013), 'Aktivisty Greenpeace zabralis na platformu Prirazlomnaya', accessed 15 October 2018 at http://www.greenpeace.org/russia/ru/news/2013/18-09-action-on-Prirazlomnaya/.

Interfax (2019), 'Strategiya Gazprom Nefti do 2030 goda: Rasti v dobyche bystree rynka i stat etalonom otrasli', accessed 20 December 2019 at https://www.interfax.ru/business/656140.

Knyazev, M. (2010), 'Roman Abramovich i ego pomoschnik', accessed 15 October 2018 at http://www.forbes.ru/ekonomika/lyudi/54248-roman-abramovich-i-ego-pomoshchnik.

Korostikov, M. and P. Tarasenko (2019), 'V Venezuele udvoilos chislo presidentov', accessed 26 January 2019 at https://www.kommersant.ru/doc/3861496?from=doc_mail.

Kozlov, D. (2016), 'Gazprom Neft stavit na novuyu neft', accessed 15 October 2018 at https://www.kommersant.ru/doc/3012559.

Kozlov, D. (2017a), 'S peresokrascheniem plana', accessed 16 October 2018 at https://www.kommersant.ru/doc/3208080.

Kozlov, D. (2017b), 'Gazprom Neft nashla neft na shelfe', accessed 16 October 2018 at https://www.kommersant.ru/doc/3429161.

Kozlov, D. (2018a), 'Gazprom Neft derzhitsya za lgoty', accessed 15 October 2018 at https://www.kommersant.ru/doc/3542454.

Kozlov, D. (2018b), 'Arabskie investory nashli dobychu v Rossii', accessed 15 October 2018 at https://www.kommersant.ru/doc/3549668.

Landes, A. and A. Nagaria (2003), 'Russian oil and gas: Taking stock', *Renaissance Capital Research*, 1–2.

Lenta.ru (2003), 'ExxonMobil vedet peregovory o pokupke kontrolnogo paketa YUKOS-Sibnefti', accessed 23 January 2019 at https://lenta.ru/news/2003/10/03/exxon/.

Lysova, T. (2005), 'Kompaniya nedeli: Puteshestviye Sibnefti', accessed 15 October 2018 at https://www.vedomosti.ru/newspaper/article.shtml?2005/09/29/97641.

Malkova, I. and E. Mazneva (2007), 'Neft na dvoikh', accessed 15 October 2018 at https://www.vedomosti.ru/newspaper/articles/2007/12/28/neft-na-dvoih.

Malkova, I. and I. Reznik (2006), 'Kto upravlyaet Gazpromom', accessed 15 October 2018 at https://www.vedomosti.ru/newspaper/articles/2006/11/22/kto-upravlyaet -gazpromom.

Mazneva, E. (2007), 'Gazpromu ne khvataet deneg', accessed 15 October 2018 at https://www.vedomosti.ru/newspaper/articles/2007/11/09/gazpromu-ne-hvataet -deneg.

Mazneva, E. (2009), 'Snova lider', accessed 16 October 2018 at https://www.vedomosti .ru/newspaper/articles/2009/04/16/snova-lider.

Mazneva, E. (2010a), 'Gazprom Neft zakhvatila somnitelnoe liderstvo – po zatratam na dobychu', accessed 15 October 2018 at https://www.vedomosti.ru/business/articles/ 2010/08/19/spornoe_liderstvo.

Mazneva, E. (2010b), 'Tsena Sibir Energy', accessed 15 October 2018 at https://www .vedomosti.ru/newspaper/articles/2010/05/26/cena-sibir-energy.

Mazneva, E. (2011), 'Khozyaika Sibir Energy', accessed 15 October 2018 at https:// www.vedomosti.ru/newspaper/articles/2011/02/16/hozyajka_sibir_energy.

Mazneva, E. (2012), 'Gazprom Neft mozhet kupit Albpetrol', accessed 16 October 2018 at https://www.vedomosti.ru/business/articles/2012/07/18/gazprom_idet_v _albaniyu.

Melnikov, K. and A. Gabuev (2011), 'Gazprom Nefti nashli dublera', accessed 16 October 2018 at https://centrasia.org/newsA.php?st=1318565640.

Melnikov, K., E. Kuznetsova and A. Gudkov (2012), 'Grecheskiy benzin ischet rossi-yskikh pokupateley', accessed 16 October 2018 at https://www.kommersant.ru/doc/ 1911591.

Mordyushenko, O. (2014), 'Gazprom Neft ne zaderzhalas v Afrike', accessed 16 October 2018 at https://www.kommersant.ru/doc/2400335.

Neft i kapital (2004a), 'Zalogovye auktsiony', *Neft i kapital*, **10**, 37.

Neft i kapital (2004b), 'Yuksi, YUKOS-Sibneft', *Neft i kapital*, **10**, 64.

Neft i kapital (2004c), 'Sibneft', *Neft i kapital*, **10**, 24.

Neft i kapital (2004d), 'Intensifikatsiya dobychi', *Neft i kapital*, **10**, 153.

Neft i kapital (2004e), 'ONACO', *Neft i kapital*, **10**, 121.

Neft, gaz i fondovyy rynok (2008), 'Sibneft', accessed 15 October 2018 at http://www .ngfr.ru/library.html?sibneft.

Neftegaz (1999), 'Otets-osnovatel Gazproma', accessed 15 October 2018 at https:// magazine.neftegaz.ru/index.php?option=com_content&task=view&id=115.

Neftegaz (2012), 'SIBUR zapustil Vyngapurovskiy GPZ, zavershiv realizatsiyu plana po pererabotke PNG', accessed 23 January 2019 at https://neftegaz.ru/news/view/ 104335-Sibur-zapustil-Vyngapurovskiy-GPZ-zavershiv-realizatsiyu-plana-po -prerabotke-PNG.

Neftegaz (2013), 'Gazprom Neft: Strategiya razvitiya do 2025 goda', accessed 23 January 2019 at https://neftegaz.ru/news/view/110240-Gazprom-neft.-Strategiya -razvitiya-do-2025-g.

Neftegazovaya vertikal (2006a), 'Perviy milliard dlya Gazproma', *Neftegazovaya vertikal*, **13**, 42–4.

Neftegazovaya vertikal (2006b), 'O roli lichnosti v neftedobyche', *Neftegazovaya vertikal*, **18**, 4–6.

Neftegazovaya vertikal (2015), 'Sanktsii na puti k 100 millionam', *Neftegazovaya vertikal*, **2**, 34–40.

Neftegazovaya vertikal (2016), 'Gazprom Neft otkazalas ot pokupki 49% edinstvennogo vietnamskogo NPZ', accessed 16 October 2018 at http://www.ngv.ru/news/gazprom_neft_otkazalas_ot_pokupki_49_edinstvennogo_vetnamskogo_npz_/?sphrase_id=292458.

Neftianka (2018), 'Gazprom Neft pozvala Shell na Ayashskiy', accessed 23 January 2019 at http://neftianka.ru/gazprom-neft-pozvala-shell-na-ayashskij/.

NEWSru (2004), 'Experty schitayut pokazukhoi pretenzii nalogovikov k Sibnefti', accessed 15 October 2018 at https://www.newsru.com/finance/03mar2004/sibneft.html.

Nikolaeva, A. (2005), 'Odin za vsekh oligarchov', accessed 16 October 2018 at https://www.vedomosti.ru/newspaper/articles/2005/09/13/odin-za-vseh-oligarhov.

Peretolchina, A., A. Fialko and A. Grechman (2010), 'Zaburitsya v Gvineyu', accessed 16 October 2018 at https://www.vedomosti.ru/newspaper/articles/2010/06/28/zaburitsya-v-gvineyu.

Petlevoy, V. (2017), 'Gazprom Neft zovet partnerov v Afriku', accessed 16 October 2018 at https://www.vedomosti.ru/business/articles/2017/03/29/683288-gazprom-neft.

Petlevoy, V., F. Sterkin and A. Toporkov (2018), 'Gazprom Neft sozdaet sobstvennyi rezervnyi fond', accessed 16 October 2018 at https://www.vedomosti.ru/business/articles/2018/02/12/750608-gazprom-neft-rezervnii-fond.

Petrachkova, A. and E. Derbilova (2007), 'Ne podelili Tomskneft', accessed 15 October 2018 at https://www.vedomosti.ru/newspaper/articles/2007/07/13/ne-podelili-tomskneft.

Podobedova, L. (2015), 'Gazprom Neft poluchila rekordnuyu otsrochku po dobyche nefti na shelfe', accessed 16 October 2018 at https://www.rbc.ru/business/15/11/2015/5645a9429a7947c868dcadf9.

Proskurnina, O. (2005), 'Grazhdanin Chukotki', accessed 16 October 2018 at https://www.vedomosti.ru/newspaper/articles/2005/10/24/grazhdanin-chukotki.

Raibman, N. (2011), 'Iran lishil Gazprom Neft prava na razrabotku mestorozhdeniya Azar', accessed 16 October 2018 at https://www.vedomosti.ru/business/articles/2011/08/29/iran_lishil_gazmprom_neft_prava_na_razrabotku.

RBC (2005), 'Gazprom pokupaet 72.6% aktsiy Sibnefti za $ 13 mlrd', accessed 24 January 2019 at https://www.rbc.ru/economics/28/09/2005/5703c2659a7947dde8e09d60.

Reznik, I. (2006a), 'Sibneft ne podkachaet', accessed 15 October 2018 at https://www.vedomosti.ru/newspaper/articles/2006/02/10/sibneft-ne-podkachaet.

Reznik, I. (2006b), 'Gazprom Neft ne uidet s birzhi', accessed 15 October 2018 at https://www.vedomosti.ru/newspaper/articles/2006/09/29/gazprom-neft-ne-ujdet-s-birzhi.

Reznik, I. (2006c), 'Intervyu: Alexander Ryazanov, president Sibnefti', accessed 15 October 2018 at https://www.vedomosti.ru/newspaper/articles/2006/02/15/intervyu-alexandr-ryazanov-prezident-sibnefti.

Reznik, I. (2006d), 'Monopoliya torguetsya', accessed 15 October 2018 at https://www.vedomosti.ru/newspaper/articles/2006/07/25/monopoliya-torguetsya.

Reznik, I. and E. Derbilova (2005a), 'Sibir vybivaet Sibneft', accessed 15 October 2018 at https://www.vedomosti.ru/newspaper/articles/2005/08/31/sibir-vybivaet-sibneft.

Reznik, I. and E. Derbilova (2005b), 'Nachalnik nashelsya', accessed 15 October 2018 at https://www.vedomosti.ru/newspaper/articles/2005/10/19/nachalnik-nashelsya.

Reznik, I. and A. Temkin (2006), 'Sibnefti skoro ne stanet', accessed 15 October 2018 at https://www.vedomosti.ru/newspaper/articles/2006/04/03/sibnefti-skoro-ne -stanet.

Reznik, I. and A. Tutushkin (2006), 'Gazprom Neft ne pustili na MNPZ', accessed 15 October 2018 at https://www.vedomosti.ru/newspaper/articles/2006/06/23/gazprom -neft-ne-pustili-na-mnpz.

Rozhkova, M. and I. Reznik (2009), 'Khroniki 1999–2009: Skvazhina Abramovicha', accessed 15 October 2018 at https://www.vedomosti.ru/newspaper/article.shtml ?2009/07/20/205787.

Salomon Brothers (1996), *Russian Oil, Vertically Integrated Companies*, Volume 3 of Report on Russian Oil, March, London: Salomon Brothers.

Savushkin, S. (1998) 'Ne sliyaniye, tak pogloscheniye', *Neft i kapital*, **25** (8), 25–6.

Serov, M. and M. Chelpanova (2013), 'Goskompanii uekhali vo Vietnam', accessed 16 October 2018 at https://www.vedomosti.ru/newspaper/articles/2013/11/13/ goskompanii-uehali-vo-vetnam.

SeverPress (2014), 'Novy variant granits Messoyakhinskogo zakaznika utverdyat na obshchestvennykh slushaniyakh', accessed 23 January 2019 at https://sever-press .ru/2014/09/15/novyj-variant-granits-messoyakhinskogo-zakaznika-utverdyat-na -obshchestvennykh-slushaniyakh/.

Solodovnikova, A. (2013a), 'Gazprom Neft razlivaetsya po Kurdistanu', accessed 16 October 2018 at https://www.kommersant.ru/doc/2135917.

Solodovnikova, A. (2013b), 'Gazprom Neft vyshla s Kuby posukhu', accessed 16 October 2018 at https://www.kommersant.ru/doc/2279921.

Starinskaya, G. (2016), 'Gazprom Neft ne hochet na shelf', accessed 16 October 2018 at https://www.vedomosti.ru/business/articles/2016/03/25/635063-gazprom-neft -hochet-na-shelf.

Starinskaya, G. (2018), 'Gazprom Neft voshla v troiku krupneishikh proizvoditeley nefti v Rossii', accessed 16 October 2018 at https://www.vedomosti.ru/business/ articles/2018/02/15/751116-gazprom-neft-voshla-troiku-krupneishih.

Surzhenko, V. and I. Reznik (2006), 'Tsena pisma', accessed 15 October 2018 at https://www.vedomosti.ru/newspaper/articles/2006/07/24/cena-pisma.

TRINFICO (1996), 'Sibneft: Investigation of investment attractiveness', Internal Report by TRINFICO, Moscow.

Tutushkin, A. (2005), 'Novaya pobeda Sibnefti', accessed 15 October 2018 at https:// www.vedomosti.ru/newspaper/articles/2005/08/25/novaya-pobeda-sibnefti.

Tutushkin, A. (2006), 'Gazprom Neft zapravit Kirgiziyu', accessed 16 October 2018 at https://www.vedomosti.ru/newspaper/articles/2006/06/21/gazprom-neft-zapravit -kirgiziyu.

Tutushkin, A. (2007), 'Gazprom Neft vybiraetsya iz yamy', accessed 15 October 2018 at https://www.vedomosti.ru/newspaper/articles/2007/06/22/gazprom-neft -vybiraetsya-iz-yamy.

Tutushkin, A. (2008a), 'Gazprom Neft prishla s mirom', accessed 15 October 2018 at https://www.vedomosti.ru/newspaper/articles/2008/02/21/gazprom-neft-prishla-s -mirom.

Tutushkin, A. (2008b), 'Doshla do Serbii', accessed 16 October 2018 at https://www .vedomosti.ru/newspaper/articles/2008/12/25/doshla-do-serbii.

Tutushkin, A. (2009a), 'Gazprom Neft uglublyaetsya v Iran', accessed 16 October 2018 at https://www.vedomosti.ru/business/articles/2009/11/12/gazprom-neft -uglublyaetsya-v-iran.

Tutushkin, A. (2009b), 'Voshla vo vkus', accessed 16 October 2018 at https://www .vedomosti.ru/newspaper/articles/2009/04/23/voshla-vo-vkus.

Tutushkin, A. and I. Vasilev (2010), 'Prishla v Kazakhstan', accessed 16 October 2018 at https://www.vedomosti.ru/newspaper/articles/2010/08/05/prishla-v-kazahstan.

Vakulenko, S. (2018), 'Gazprom Neft: Osnovnye idei strategicheskogo razvitiya', accessed 8 February 2019 at https://www.gazprom-neft.ru/press-center/lib/1642212/ ?sphrase_id=5595871.

Vinogradov, I. (1998), 'Elf tochno otkazalas ot Rossii', accessed 24 January 2019 at https://www.kommersant.ru/doc/204019.

Zanina, A. (2011), 'Ya diplom ne pisal. Vsyo ostalnoe delal', accessed 15 October 2018 at https://www.kommersant.ru/doc/1807636.

5.   Surgutneftegas: quiet conservative

Surgutneftegas's response to changing times has been to resist change as much as possible through a strategy of tight control. Of all the heads of the Soviet oil enterprises, only Surgutneftegas Director Vladimir Bogdanov and Vagit Alekperov of LUKOIL have succeeded in defeating the oligarchs and retaining control of the companies (see the chapter on LUKOIL in this book).

Surgutneftegas is considered to be Russia's most conservative, least transparent and perhaps most financially stable oil company. It has always paid its taxes promptly, even when tax evasion was rife in the country. *Forbes* magazine stated that 'as a competent and honest industrial tycoon, Bogdanov stands out in Russia' (Forbes 2001). Surgutneftegas has also supported other domestic industries, kept politics and foreigners at arm's length and demonstrated solid organic growth (Poussenkova 2004). Bogdanov has remained a loyal and respected citizen of Khanty-Mansi Autonomous District, where most of the company's exploration and production has been concentrated, although the company faces protests from some of the indigenous peoples displaced by Surgutneftegas activities.

Surgutneftegas has proved to be resilient to the challenges presented by low oil prices, financial crises and Western sanctions. It is one of the few oil companies in Russia that can produce hard-to-recover reserves using its own technology and expertise without international partners. While the company does not make grand statements on climate change, it has employed some relatively climate-friendly, efficient technologies, and its level of utilization of associated petroleum gas (APG) has consistently been the highest in Russia. Shvarts et al. (2015), Shvarts et al. (2016) and Shvarts et al. (2018) rate Surgutneftegas as the most environmentally responsible Russian oil company.

With oil production of 61 million tonnes in 2018, Surgutneftegas is the fourth-largest Russian oil company after Rosneft, LUKOIL and Gazprom Neft. Unlike most of the Russian oil companies, it is headquartered near its main assets, in the town of Surgut in Western Siberia. The company is wholly privately owned, although its precise shareholder structure remains unknown.

The company's exploration and production operations are concentrated in Western Siberia, Eastern Siberia and Timan-Pechora. Surgutneftegas owns the Kirishinefteorgsyntez (KINEF) refinery, located in Leningrad region near Saint Petersburg, and five petroleum product distributors in the same region: Kaliningradnefteproduct, Kirishiavtoservis, Novgorodnefteproduct,

Pskovnefteproduct and Tvernefteproduct (295 fuel stations in total) as well as the Surgut gas-processing plant and two R&D institutes (SurgutNIPIneft, based in Surgut with a branch in Tyumen) and Lengiproneftekhim (Saint Petersburg).

CORPORATE HISTORY

The predecessor of the present-day company, Surgutneftegas Production Association, was established in 1977. This was a decision-making body with oversight over several state oil-producing enterprises. Vladimir Bogdanov, aged 33, was appointed General Director of Surgutneftegas Production Association in 1984, becoming the youngest director ever of the largest West Siberian production association.

The 1990s: Unchanging Surgutneftegas in a Changing Russia

Along with LUKOIL, YUKOS and Rosneft, Surgutneftegas was established by Presidential Decree 1403 of 19 November 1992 (Kremlin 1992), which enabled the creation of vertically integrated oil companies by merging exist-ing state-owned companies and assets. In 1992, Vladimir Bogdanov enjoyed the strong support of the Russian government and was able to handpick the assets he wanted to incorporate into Surgutneftegas, including the KINEF refinery. Analysts observed that among the oil companies that appeared in the 1990s, Surgutneftegas had the tightest fit between its subsidiaries and the most streamlined corporate structure. The open joint stock company (OJSC) Surgutneftegas Oil Company was established by the Russian gov-ernment's order 271 of 19 March 1993, 'On the Establishment of the OJSC Surgutneftegas Oil Company'.

Surgutneftegas was one of the first vertically integrated oil companies in Russia to be privatized. Vladimir Bogdanov was determined from the outset to establish firm control over the company, and he accomplished this goal. In 1993, 8% of Surgutneftegas shares were sold by the federal government at a closed auction. Surgutneftegas purchased another 7% of its own shares for vouchers. At this time, 45% of the shares remained in the hands of the federal government. In June 1994, Khanty-Mansi Autonomous District sold 40% of Surgutneftegas shares via a tender to the company Neft-Invest, which was controlled by Surgutneftegas (Nefte.ru 2019).

Bogdanov decided to arrange a loans-for-shares auction for Surgutneftegas along the lines of other loans-for-shares auctions that took place in the early 1990s. He managed to limit access to the auction: Rosneft, for instance, was excluded on a technicality (Gustafson 2012). At that time, Bogdanov had a powerful ally in Moscow, Vladimir Potanin, founder of ONEXIM Bank and a key player in the wider loans-for-shares scheme. Thus, in November 1995, 40.12% of the Surgutneftegas shares that were still being held by the state were sold at the loans-for-shares auction. The Surgutneftegas Pension Fund, established by the company's management, received this stake in trust by pledging to pay the government the tax debt of Surgutneftegas Oil Company (amounting to USD 216 million) and provide the state with a loan worth USD 84 million. Potanin agreed that ONEXIM Bank would act as a guarantor of the Surgutneftegas Pension Fund at the auction. In return, Surgutneftegas kept its accounts in ONEXIM Bank for several years (Forbes 2004). Surgutneftegas also bought a 16% share in ONEXIM Bank, partly as a good investment and partly to keep an eye on its partner.

When the stake was resold in 1997, the relevant obligations included a provision of financing for Surgutneftegas production programmes (RUB 1.1 trillion for three years) as well as local social and economic development programmes (RUB 100 billion for one year). The stake was won by Surgutfondinvest, a previously unknown pension fund manager, which turned out to be a member of the Surgutneftegas family (The Moscow Times 1997). The Surgutneftegas privatization was complete.

In the mid-1990s, Bogdanov also focused on tightening his control over the exports of Surgutneftegas oil. He found a trading partner in the company Nafta-Moscow, the former Soviet Soyuznefteexport, and bought 15% of its shares in 1997, thus ensuring a stable channel for exports of its oil. In 2001, when Nafta began to lose its market position, Bogdanov sold the shares again.

Another export-trading partner of Surgutneftegas was Kineks, owned, among others, by Gennadiy Timchenko. This alliance had long-term strategic implications for both partners. Kineks was established in the mid-1990s and began to flourish in the late 1990s when it was already exporting 60% of Surgutneftegas's petroleum products and had begun to tighten its grip on exports of its crude, gradually displacing Nafta-Moscow. In 2003, Timchenko broke up with his partners in Kineks and established the oil-trading company Surguteks, where he had managed to capture 51% of Kineks's business (Vedomosti 2012a). Another oil-trading company, Gunvor, was established by Timchenko in 1997, and also became an important trader of Surgutneftegas's oil and petroleum products.

Surgutneftegas's leaders – in effect, mainly engineers by background – showed considerable sophistication in financial matters. Thus, in early 1995, Surgutneftegas was the first Russian company to make an additional issue of shares. It also issued first-level American depository receipts (ADRs) in 1997, becoming the fourth Russian oil company after LUKOIL, Chernogorneft and Tatneft to enter Western stock markets.

Surgutneftegas differed from other vertically integrated Russian oil companies in that it never used tax-optimization schemes. It was always one of the most punctual taxpayers in Khanty-Mansi Autonomous District and had the firm support of the local authorities (Kommersant 2000). Moreover, Surgutneftegas has never had wage arrears, while its salaries are higher than those paid by many other oil companies.

Probably because of this fiscal discipline, Surgutneftegas successfully lobbied for its interests with the federal government and often obtained more benefits than its competitors. For instance, in 1996, the government permitted the company to export an additional 10 million tonnes of oil for two years to finance the construction of an export terminal in Batareynaya Bay in the Gulf of Finland; Surgutneftegas was allowed to pump the whole allocated volume while the export schedules of its rivals were curtailed.

During the 1990s, there was an avalanche of new joint ventures between foreign companies and Russia's oil majors (Overland et al. 2013). Yet during this time, Surgutneftegas was the only major that did not establish joint ventures with foreigners. According to Bogdanov, 'oil produced by most joint ventures could be produced by domestic enterprises using our technology and would cost the Russian side much less' (Russian Petroleum Investor 1997). There may have been another reason why other Russian oil companies opted for joint ventures while Surgutneftegas avoided them. They often regarded them as vehicles for increasing export volumes; however, Surgutneftegas was already exporting more crude oil than its Russian peers.

Although his power has been challenged in different ways, Vladimir Bogdanov has suffered surprisingly few defeats over the decades. One such challenge came from the indigenous Khanty people living close to the Tyanskii reserves in Khanty-Mansi Autonomous District, who feared that oil development threatened their reindeer herding and sacred sites. In March 1991, the Khanty of Russkinskii village held a referendum and voted unanimously against oil extraction in their local area. The project received a negative review from the state ecological experts; however, this decision was overturned by the federal government, which resulted in so many local protests that the district government was forced to announce a moratorium on the oil developments until 1996. In fact, Surgutneftegas resumed work on the Tyanskii field in 1994, but the deal that came out of this conflict established a precedent for engagement between indigenous peoples and oil companies in the Russian North, including the negotiation of company-community agreements.

Bogdanov's most significant failure during the 1990s was probably his unsuccessful attempt at controlling the petroleum products distribution in Saint Petersburg, which resulted in a major fuel shortage in the city in December 1994 and the restructuring of Surgutneftegas as a business. Following Bogdanov's efforts in 1994 to control distribution through strong-arm tactics and share swaps, the Saint Petersburg authorities set up a competing company, and Surgutneftegas eventually lost control over its distribution network. The situation was compounded when a local gang boss took control of the local market. In 1998, Surgutneftegas sold off the last of Surgutneftegas's distribution and retail assets in Saint Petersburg, although it retained some distributors in the wider Leningrad region (Gustafson 2012).

The 2000s

While Putin's Russia differed radically from Yeltsin's Russia, Surgutneftegas stuck to the strategy and management principles it had developed during the 1990s. Strengthening a grip over the company was still one of the key objectives of its leadership. Surgutneftegas has maintained good relations with the political elite throughout the Putin era, although it has resisted getting more involved than necessary. Bogdanov was Putin's close ally in the 2000 elections, but he subsequently declined an offer to become the Minister of Energy.

In the 2000s, support for domestic industry remained one of the top priorities for Surgutneftegas. The company formulated a strategic objective to replace 50% of imported technologies and materials with Russian analogues. In particular, it cooperated with the former defence enterprises to convert them to manufacturing of oilfield equipment (Burneft.ru 2017). In 2003, Surgutneftegas placed orders for 106 types of equipment and spare parts at 22 Russian enterprises. Given Surgutneftegas's large outlays for high-tech gear, it meant significant private investments in Russia's domestic R&D potential.

Surgutneftegas also demonstrated efforts towards more responsible oil development. In 2000, it became one of the few oil companies that gave up the practice of dumping wastewater into natural reservoirs, instead channelling them back into the oil and gas fields in order to maintain reservoir pressure.

In the new millennium, Surgutneftegas has managed to achieve considerable success in its E&P despite the limitations of the ageing reserve base (see the section on production strategy below). However, its track record of downstream operations is less impressive. In June 2002, Surgutneftegas, Rosneft and the government of Leningrad region signed an agreement on building a refinery and a sea terminal in Primorsk at the cost of USD 1.3 billion to process 8–10 million tonnes per annum. However, in 2004, Rosneft withdrew from the project, and in 2006, Surgutneftegas also gave up its plans, with Bogdanov stating that the shortage of electricity and water had made the construction commercially unfeasible (Vedomosti 2006). Also, in 2013, Surgutneftegas terminated the construction of the oil-loading terminal at Batareynaya Bay, a project it had been half-heartedly implementing since the late 1990s (NIAR 2014).

The 2000s were a time for further tightening of Bogdanov's grip on Surgutneftegas. In the year 2000, Bogdanov consolidated his holding in an unusual manner. Shares of JSC Surgutneftegas, the oil-producing subsidiary, were swapped for shares of the parent company, Surgutneftegas Oil Company, and the refinery and petroleum product subsidiaries. Initially, many small-scale shareholders viewed this consolidation negatively, fearing the dilution of their stock. However, the share swap ratios proposed by the company seemed acceptable to everybody, and consolidation was achieved without scandals or lawsuits (Newsruss.ru 2019).

In 2003, Bogdanov further consolidated his control over the company. After the TNK-BP alliance and failure of the YUKOS-Sibneft deal, everybody was expecting new mergers and acquisitions (M&A) in the sector and considered Surgutneftegas to be the prime candidate for a hostile takeover by Sibneft or TNK. In spring 2003, there were indications that the Sovlink brokerage firm acting for TNK was actively purchasing the stock of Surgutneftegas. In a counteroffensive, Surgutneftegas, without much regard for price, aggressively bought a block of its own stock sufficient to ensure a controlling interest for Bogdanov. As a precautionary measure, Surgutneftegas Oil Company, the key subsidiary that held the most significant block of Surgutneftegas shares (36.7% of authorized capital and 42% of voting stock), was renamed Leasing Production Ltd and was transformed from an OJSC into a limited liability company, making it even less transparent (Neft i kapital 2004).

In late 2006, Surgutneftegas sold the structure called Leasing Production. Analysts estimated the value of the deal at USD 20 billion, one of the biggest transactions carried out in Russia to that date. However, it was unclear who bought this stake (external buyers or Surgutneftegas entities) and why Surgutneftegas initiated this deal (probably as a precautionary measure against a hostile takeover) (Kompromat 2007). Subsequently, in early 2007, information was leaked indicating that the Surgutneftegas Pension Fund had become the owner of Leasing Production (Vedomosti 2007b).

The same year, that is, 2007, analysts inferred that 72% of Surgutneftegas shares (85% of voting stock) might belong to 23 non-commercial partnerships that kept them under the heading of long-term financial investments in their accounts. These partnerships were either established by Surgutneftegas or controlled by its top managers, including Bogdanov (Vedomosti 2007b); however, the ownership of these partnerships was later transferred to ordinary employees of Surgutneftegas.

In the new millennium, Surgutneftegas has continued to maintain its reputation as an investor-unfriendly company – a reputation originally established through its practice of limiting investor access to share auctions and engaging in share-swapping to consolidate control over the company. Minority shareholders have frequently expressed dissatisfaction about Bogdanov's dividend policy. In 2001, the international investment fund Russian Prosperity sued Surgutneftegas for its practice of calculating the profit that allegedly led to the under-statement of dividends on preferred shares but lost the case.

A further disagreement in 2004 was connected with a lawsuit by Harvard University Pension Fund, which held 3 million ADRs on preferred shares. The Fund claimed that over six years, Surgutneftegas had intentionally declared dividends far below the amount mandated by the company's charter; it had managed to do so by using an artificially under-stated net profit figure by deducting from revenues not only operational expenses and taxes but also investments. Since management controls the bulk of Surgutneftegas's common shares, the plaintiff argued that Surgutneftegas's retention of virtually all the company's earnings benefited its executives at the expense of holders of preferred shares. The Harvard University Pension Fund estimated its 2003 losses at USD 3.7 million (Ria Novosti 2004).

In 2004 and 2005, William Browder, Head of the Hermitage Capital hedge fund, initiated a lawsuit against Surgutneftegas, demanding redemption of its treasury stock (the common stock owned by Surgutneftegas itself) but lost the case (Gazeta.ru 2013). Browder has referred to Bogdanov as 'the Saddam Hussein of Russian business' (Bloomberg 2015). In 2005, Browder's Russian visa was annulled.

Surgutneftegas continued to be one of the most closed and least transparent Russian oil companies. In fact, information about the key shareholders remains one of the best kept secrets in the Russian oil industry. There were many rumours about the owners, one being that Gennady Timchenko held some 25% of the company. However, Timchenko denied that he owned a large stake in Surgutneftegas, and his representative told the *Financial Times* in 2013 that he held less than 0.01% of the company's stock (BBC 2013).

During the 2000s, there were also unconfirmed rumours that Vladimir Putin owned 37% of Surgutneftegas, although the Kremlin vehemently denied it (Ino TV 2014). Given these rumours, it is unsurprising that Surgutneftegas managed to avoid being acquired by other Russian oil companies, although the market was regularly kept on edge by speculations about a potential purchase of Surgutneftegas by Gazprom or Rosneft.

However, one important characteristic of Surgutneftegas did change during the 2000s: the formerly apolitical company was gradually drawn into the political agenda of the new Russian leadership. Vladimir Putin and Igor Sechin likely became acquainted with Bogdanov in the early 1990s when they both worked in Saint Petersburg mayor's office (Znak 2015). It is noteworthy that the title of Igor Sechin's PhD (kandidatskaya) thesis, which he defended at Saint Petersburg Mining University in 1998, was 'The economic valuation of investment projects of oil and petroleum products transit: the case of the Kirishi-Batareynaya petroleum product pipeline' (Lenta.ru 2019). This pipeline is owned by Surgutneftegas.

Throughout the 2000s, Surgutneftegas was increasingly associated with the *siloviki*[1] group (politicians from the military or security services) in the Russian establishment, and especially with Igor Sechin, who was then the Deputy Head of the Presidential Administration, and it seems that the company was sometimes asked to perform certain services for the Kremlin. For instance, during the 2004 auctioning of a former key asset of YUKOS, Yuganskneftegaz, Baikal Finance Group appeared out of nowhere and submitted the winning bid. Baikal Finance Group was represented at the auction by two managers from Surgutneftegas, Igor Minibayev, Head of the Department of Organizational Structures, and Valentina Komarova, First Deputy Head of the Financial Department (Vedomosti 2014).

Similarly, in 2005, Surgutneftegas bought 35% of the Ren TV channel from Severstal for between USD 60 and 100 million, according to different estimates. This was surprising as Surgutneftegas had previously avoided the acquisition of non-core assets. In fact, this was one of the steps taken by the state to tighten its control over the mass media by bringing key media organizations under the control of loyal companies. According to newspaper reports, it may have been Sechin who proposed the idea of selling a stake in Ren TV to Surgutneftegas (Vedomosti 2005a). Continuing this trend, in March 2009, Surgutneftegas raised its stake in National Media Group from 12.3% to 24%. The group owns 51% of National Telecommunications, 100% of Ren TV, 72% of the Petersburg Channel 5 and 51% of *Izvestiya* newspaper (Kommersant 2009).

This close association with politics poses a certain risk to Surgutneftegas in that it could be asked to do even more serious favours for the Kremlin. For instance, when a 19.5% stake of Rosneft was to be sold off in the second round of its privatization, speculation was rife that Surgutneftegas would be the buyer and that its funds would be used to cover USD 23.5 billion of Rosneft's debt, although this was never confirmed (Ihodl.com 2015).

Post-2010

In 2012, Surgutneftegas changed its ownership structure: 13 of the 23 non-commercial partnerships mentioned above ceased to exist and were merged with other partnerships. Their owners remained practically the same; however, there was a significant decrease in the total financial investments of the new non-commercial partnerships. This reorganization, which further complicated the already very complex ownership structure, might have been connected with Surgutneftegas's pending publication of its financial statements under international accounting standards (IAS) for the first time in ten years as a way for Surgutneftegas to avoid disclosing information about its owners (Kommersant 2012a) (see the section on transparency below).

Surgutneftegas continued to have only the one refinery that it obtained during privatization. KINEF is located in Leningrad region and produces over

80 different petroleum products. Surgutneftegas has continued to upgrade KINEF over the years.

Surgutneftegas still was able to obtain significant benefits from the government through its alliances with politically influential partners. Thus, in 2012, the Russian oil companies found themselves unable to upgrade their refineries in time to comply with the terms established by the government in October 2011 and asked the ministry to extend the deadline. However, at that time, only Surgutneftegas managed to reach an agreement with the officials concerning such an extension. The company was allowed to postpone the commissioning of isomerization and catalytic reforming installations from 2015 to 2016, hydrocracking from 2012 to 2013 and hydro-purification from 2014 to 2015 (Kommersant 2012b).

In June 2017, Vladimir Putin signed a Decree on State Awards. In recognition of Surgutneftegas's contribution to Russia's innovation potential, Bogdanov and his deputies Yuri Baturin and Anatoliy Nuryaev received science and technology awards for creating a sustainable system of development for the petroleum fields of Western Siberia (Kommersant 2017).

COMPANY PROFILE

This section covers key aspects of Surgutneftegas's operations, which help set out the company's corporate profile. These include the company's production strategy, involvement in the Arctic and in offshore oil and gas developments, internationalization strategy, operational transparency and corporate social responsibility.

Production Strategy

The production strategy developed by Surgutneftegas during the 1990s was quite unusual for Russia. The company aimed at stabilizing output based on its existing assets rather than through acquisitions. It targeted exploration drilling to ensure adequate reserve replacement and was the only oil company in Russia to establish an exploration division in 1995. It brought on stream several new fields (for example, Tyanskoye and Konitlorskoye with reserves of 300 million tonnes in 1994). One analyst observed in 1996 that Surgutneftegas was 'one of the few companies thinking about the future' while its competitors 'were living off the fields discovered in the Soviet times with state funds when no one counted the money being invested in the sector's development' (Russian Petroleum Investor 1997, p. 70). As a result of this strategy, Surgutneftegas's production decline from 1994 to 1995 (2.4% per annum) was much lower than the sector average (8–10% per annum). It is also noteworthy that

Surgutneftegas's oil output had already begun to grow in 1996 while Russia's other vertically integrated oil companies posted growth only in 2000.

However, Surgutneftegas's rates of growth began to decline in the mid-2000s: in 2003, production grew by 11.7%, and 7.11% and 2.65% in 2005 and 2006, respectively. Some analysts believed that this slowdown happened because of the heavier financial burden on the sector, which was not being compensated for by the rising oil prices. Others thought that it resulted from the natural depletion of Surgutneftegas's West Siberian fields with no major new acquisitions to offset the decline. Some drew on horizontal drilling to explain this phenomenon, which the company mastered in 2001 and 2002, and argued that Surgutneftegas leadership relied excessively on these aggressive methods of field development (Vedomosti 2007a).

From 2007 onwards, Surgutneftegas's production declined for five consecutive years, and only in 2011 did it increase again by 2.1% due to the commissioning of new fields – Talakansk and Alinsk – in the Republic of Sakha (Yakutiya) (see below). In addition, the company gradually managed to bring the mature fields' rates of decline under control: in 2006, the rate was 6.2% and 2.8% and 1.8% in 2010 and in 2011, respectively. This was accomplished by intensifying its drilling programme (Kommersant 2012c).

Surgutneftegas has always focused on increasing its oil recovery ratio, which at 0.43 is one of the highest among Russian oil companies (Neft i kapital 2012). Annually, it undertakes over 9000 operations to enhance oil recovery. The company has also always had the lowest number of idle wells in the total well count among Russian oil companies. The company's APG utilization has also consistently been the highest in Russia – 99.56% in 2018 (Surgutneftegas 2018) (see the section on climate change below).

Surgutneftegas's E&P activities have always been concentrated primarily in Khanty-Mansi Autonomous District, and the company stayed away from the 'asset-grabbing fever' in Russia in the mid-1990s, avoiding growth for the sake of growth. However, by the mid-2000s, as the limitations of the ageing West Siberian fields became increasingly visible, Bogdanov changed his approach. In 2003, Surgutneftegas shed its reputation of a Khanty-Mansi recluse by gaining the rights to the Talakansk field, with 124 million tonnes of reserves in the Republic of Sakha (Yakutiya). Initially, Sakhaneftegaz, controlled by YUKOS, had won the licence for Talakan in 2001. The results were subsequently annulled because Sakhaneftegaz was unable to pay the required bonus of USD 501 million. Another investment contest was planned; however, in 2002, Surgutneftegas, as the bidder that offered the second-largest bonus (USD 61 million), asked the arbitration court of Sakha to recognize it as the winner. In late 2003, it was decided that the licence be granted to Surgutneftegas, and in April 2004, Surgutneftegas and the Sakha Republic signed an agreement for the development of the field. Surgutneftegas's growing oil production in

Eastern Siberia reached 9.1 million tonnes in 2018, with the region accounting for 15% of the total output of the company (Surgutneftegas 2018).

Surgutneftegas has consistently been one of the leaders in exploration and development drilling in Russia. In 2018, it accounted for 20% of Russia's total exploration drilling and 18% of development drilling. By the end of 2018, it had a total of 24 457 active wells, while the proportion of idle wells had fallen to 8%, the lowest in the sector (Surgutneftegas 2018).

Since the company rarely makes presentations for analysts, the main source of information about its production strategy is Bogdanov's speeches. At the 2012 annual general meeting (AGM) of shareholders, Bogdanov told journalists that Surgutneftegas did not plan to buy any assets: 'I do not need to buy anything. We have everything!' He also stated that the company's production strategy remained the same: 'to drill, drill, drill and develop what we [already] possess' (Bogdanov, cited in Vedomosti 2012b). At the 2013 AGM, Bogdanov repeated that the company would not change its strategy: it would develop in Russia and continue to accumulate money in bank deposits. He stated that Surgutneftegas had stabilized production at the 2012 level of 61.4 million tonnes, and crude output would grow again in the coming years mainly due to new projects, the largest among them being the Severo-Rogozhinskoye field in Western Siberia where production would begin in 2015, six years ahead of the deadline established in the licence (Kommersant 2013a). Indeed, by 2018, the Rogozhnikskaya group of five fields reached the planned production level of some 4 million tonnes of oil per annum (Surgutneftegas 2018).

Analysts have frequently said that one of the key problems for Surgutneftegas is its insufficient reserve increment (replacement of production with new reserves), which was much lower than that of LUKOIL or Rosneft. However, in December 2012, Surgutneftegas began to spend some of its financial reserves. In an intense competition with Rosneft, Gazprom Neft and Gazprombank, it bought for RUB 46 billion a production licence for the Schpielman field in Khanty-Mansi Autonomous District, containing 146 million tonnes of oil. It was Surgutneftegas's most significant purchase in recent years and the most expensive licence in Russia until then (Kommersant 2012d).

In 2016, Surgutneftegas began to develop a new area, the Uvat cluster in Tyumen region, by commissioning the Yuzhno-Nyurymskoye field with recoverable reserves of 8.5 million tonnes. In 2018, oil production at the Uvat cluster with two fields under development reached 935 000 tonnes. Within a five-year time horizon, Surgutneftegas was planning to commission 21 new fields, ensuring that production stayed at 61 million tonnes per annum.

In 2016, oil production by Surgutneftegas grew by 0.4% to 61.8 million tonnes, with East Siberian fields providing the bulk of the growth. In 2017, Surgutneftegas was to reduce its oil production to 61.1 million tonnes to contribute to the fulfilment of the requirements of the OPEC+ agreement of

November 2017, where OPEC and non-OPEC producers agreed to extend oil production cuts to the end of 2018. In 2018, production declined further to 61 million tonnes. At the same time, gas production continued to decline from 12.3 billion cubic metres in 2012 to 9.7 billion cubic metres in 2016 and remained at this level in 2018 (Surgutneftegas 2018, p. 8). The commissioning of new gas fields in the Republic of Sakha did not offset declines at the depleting fields in Western Siberia. Since there is neither a market nor an infrastructure for gas in the Republic of Sakha, Surgutneftegas had to re-inject gas, increasing the level of APG utilization in Sakha to 96.7% in 2014 (Territory neftegas 2015).

The Arctic and Offshore Development

With current operations in Western and Eastern Siberia and Timan-Pechora and a long history of activity in Khanty-Mansi Autonomous District, Surgutneftegas has ample experience of operating onshore fields in the Arctic and sub-Arctic conditions. Yet the company has never demonstrated any serious ambition to work offshore in the Arctic. In 2005, there were reports that it was negotiating the possible joint development of offshore oil and gas fields in Russia with the Norwegian oil company Statoil (later renamed Equinor) (Vedomosti 2005b). However, these plans never materialized.

In 2008, amendments were made to the Law on the Continental Shelf, restricting offshore development to projects led by Gazprom and Rosneft. Bogdanov stated that it was now only possible to undertake offshore development as a minority partner of one of the two major state companies,

> and we are not ready to work under such conditions ... We are also a decent Russian company. Our principle is the following: We fully implement the project by ourselves from the beginning to the end; we do not join partnerships; we have enough money and potential, both technological and human. (Vedomosti 2012b)

Internationalization

Over the years, Surgutneftegas made only two serious attempts to operate outside Russia. In 2008, it reluctantly joined the National Oil Consortium in Venezuela, which also included Rosneft, LUKOIL, TNK-BP and Gazprom Neft, and was established on Igor Sechin's initiative to develop the Junin-6 field. In 2012, when Sechin left his position as Deputy Prime Minister, Surgutneftegas quickly withdrew from the consortium, and Rosneft bought its stake for RUB 6 billion (Kommersant 2012e).

Another unsuccessful attempt to operate overseas was made in April 2009 when Surgutneftegas bought 21.2% of the Hungarian company MOL for EUR 1.4 billion from the Austrian company OMV. MOL considered this acquisition a hostile takeover and, as a precautionary measure, introduced amendments to its charter, significantly expanding the authority of its board of directors whereby it could limit the shareholders' rights if they did not provide information about the company's owners (or if there were reasons to doubt this information). Thus, in effect, MOL could demand to know the identity of the ultimate beneficiaries of Surgutneftegas, which refused to provide such information. Consequently, MOL did not even include Surgutneftegas in the register of its shareholders, precluding Surgutneftegas's participation in the AGM, which approved the above-mentioned amendments (Vedomosti 2009a).

In response, in June 2009, Surgutneftegas filed a lawsuit against MOL in the Court of Budapest demanding that Surgutneftegas be included in the register. It lost two court cases against MOL, which continued to deny Surgutneftegas's access to its register and shareholders' meetings. Igor Sechin, who was considered the 'main curator' of the deal, tried to help Surgutneftegas but in vain (Kommersant 2011a). Ultimately, in May 2011, Surgutneftegas sold its stake to the government of Hungary for EUR 1.88 billion, earning some EUR 500 million. Still, most Russian analysts believed that cash-rich Surgutneftegas did not need this money and that it had been requested by the Kremlin to buy a stake in MOL on behalf of one of the state companies (Kommersant 2011b).

These two failed attempts to go international reinforced Surgutneftegas's preference for operating within Russia. Indeed, in 2012, Bogdanov explained that Surgutneftegas was not interested in overseas production assets: 'Their oil is the same as ours' (Vedomosti 2012b). He reiterated this statement at Surgutneftegas's AGM in 2013: 'We believe that it is more efficient to work in Russia despite the high taxes, problems with tariffs of natural monopolies and unstable legislation ... I have not seen any information showing that other companies that work abroad earned profits that could ensure a return on investments' (Vedomosti 2013a).

Innovation

As noted above, Surgutneftegas was the only Russian oil major that decided not to establish joint ventures with foreign companies in the 1990s. Yet despite its cautious attitude towards foreigners, Surgutneftegas has been a pioneer among Russian oil companies in using Western production technologies and state-of-the-art equipment. In the mid-1990s, it was the first to implement the 'enterprise resource management' software from the leading German provider SAP to monitor and control costs. Bogdanov's approach to the new technology was cautious. However, once he had made up his mind, he sent his best people abroad by the hundreds to be trained (Gustafson 2012, p. 226). By the mid-1990s, investment analysts declared that Surgutneftegas was Russia's leader in the use of modern crude extraction technologies and one of the few

Russian oil companies capable of horizontal drilling and fracking, which it performed using in-house resources rather than drawing on the services of foreign service companies.

This focus on 'localization' was also manifested in Surgutneftegas's approach to its own scientific resources. The research institute SurgutNIPIneft initially started to use Schlumberger technology to create geological models for eight fields; however, it gradually shifted to domestic methods of digital geological and hydrodynamic modelling.

Transparency

Surgutneftegas is considered to be Russia's least transparent major oil company. This strategy has helped Bogdanov maintain control over the company. As noted above, the lack of transparency around share ownership, for example, may have reduced the risk of hostile takeovers, although it led to difficulties when attempting to acquire overseas assets.

Surgutneftegas's 2013 IAS reconfirmed the company's lack of consideration for investors. Surgutneftegas had not provided financial statements under the IAS since 2002 when it issued its 2001 IAS Report, causing a scandal. The 2001 Report stated that the treasury stock of the company accounted for 40.5% of its authorized capital or 46.6% of voting stock. The largest minority shareholders sued Surgutneftegas, demanding it not to vote for this stock. Since then, Surgutneftegas has provided statements only under Russian accounting standards. However, in 2013, it was forced to publish IAS statements under a new order by the Ministry of Finance applicable to public companies and banks (Vedomosti 2013b).

The IAS 2012 statements did not advance our knowledge of Surgutneftegas; its treasury stock and beneficial ownership remained in the dark (Vedomosti 2013c). Analysts stated that during the investor day organized on the occasion of the publication of IAS statements, Surgutneftegas confirmed its reputation as the most closed company in the sector. One analyst stated thus: 'They do not need anything. Surgutneftegas is a state within a state. The company owns practically everything in its region, and the average salary is RUB 60 000, which is much higher than regular Russian wages; they are satisfied with everything' (Kommersant 2013b).

At the same time, the IAS statements demonstrated that Surgutneftegas remained one of the main contributors of capital to the Russian banking system. In late 2012, its bank deposits exceeded the combined bank deposits of the Ministry of Finance and the regions of the Russian Federation. Surgutneftegas mainly kept its money in dollars (USD 24.3 billion) and euros (EUR 3.4 billion) in Sberbank, Gazprombank, VTB and Unicredit-Bank and earned about USD 1 billion per annum in interest (Kommersant 2013b).

Surgutneftegas publishes an annual report, which covers production, research and technology, corporate governance and securities, financial statements and key risks to business; it also reports on environmental and social issues, including expenditure in these areas and the energy resources utilized by the company. Its key performance indicators include, among others, production figures, investments, existing and new wells and gas stations, number of personnel, revenue, net profit and dividends per share. In the section on corporate responsibility, it analyses environmental safety, human resources policy, spending on social activities, charity and sponsorship. It reports on the following areas relating to environmental protection: enhancing environmental safety and prevention of accidents; atmospheric air protection; protection, rational use and re-cultivation of land; protection, rational use and restoration of water resources; waste treatment; and monitoring of the environment and production facilities (Surgutneftegas 2016a, pp. 68–70). Since 2004, Surgutneftegas has been producing an annual *Environmental Report*, which elaborates all aspects of its environmental protection activities and engagement with local and indigenous communities (Surgutneftegas 2019b).

The company's website also provides some information relating to all aspects of the company's activities, including E&P; financial reports for investors and shareholders; information for the public, including reports on environmental conflicts and community grievances; company policy concerning its relations with indigenous peoples; and information relating to protected areas, including reports and videos on Numto Nature Park.

Surgutneftegas is also listed on several foreign stock exchanges, including the London Stock Exchange. Participation in international stock exchanges obliges Surgutneftegas to adhere to minimum transparency standards.

CSR

Like other Russian vertically integrated oil companies, Surgutneftegas continues to sign and implement cooperation agreements with the administrations of the regions where it operates and enters into direct agreements with indigenous peoples (Neftegaz.ru 2008). The agreements with indigenous peoples often relate to the provision of vehicles, chainsaws, fuel, diesel generators, construction materials and so on. The company pays the transportation expenses of indigenous families, reimburses expenditure for medical treatment and care and supports children's education (Neftyaniki 2015). These arrangements have occasionally come under criticism as commitments are not always followed through, while Surgutneftegas has limited the eligibility of families to receive this kind of support, excluding households who receive income through state employment. Unlike LUKOIL, which frequently offers ad hoc assistance to indigenous peoples outside of formal agreements, Surgutneftegas is reported to have a policy that restricts assistance to formal agreements.

With its headquarters in Surgut, Surgutneftegas is one of the two vertically integrated oil companies (the other being Tatneft) that has its headquarters in a production region rather than in Moscow or Saint Petersburg. As a result, it pays considerable attention to the city, which is one of the best funded and managed in Russia. Unlike many other Russian oil companies, Surgutneftegas did not abandon the social infrastructure that it maintained during the Soviet era by transferring it to municipalities, which was a fashionable thing for oil companies to do in the 1990s. Contrary to the advice of Western consultants, the company continued to sponsor education in target areas; it also provided its staff with accommodation, medical insurance and a generous non-state lifetime pension scheme with established guaranteed payments (Ugra.mk.ru 2012).

Some of Surgutneftegas's oil production facilities are located within or in the vicinity of nature reserves and/or areas sacred to indigenous peoples, such as Numto Nature Park and the sacred Imlor Lake in Yugra (Surgutneftegas 2016a, p. 83; Vestnik 2013). Despite protests from the residents living around Imlor Lake in 2010, Surgutneftegas managed to reach an agreement with the Governor of Khanty-Mansi Autonomous District, and today all but one indigenous household have left the Imlor Lake area in tandem with advances in oil development. Greenpeace Russia has campaigned on both the Lake Imlor and Numto cases. In the case of Numto in 2015, more than 36 000 letters were sent to regional and federal officials urging them to reject proposals to open up the wetlands of Numto Nature Park to oil drilling. Public hearings were held; however, in October 2016, the Ministry of Natural Resources and Environment decided to change the zoning regulations in the park to allow oil drilling in the wetlands (Greenpeace 2017).

In 2018, Surgutneftegas spent RUB 17 billion on environmental protection – according to its own account (Surgutneftgas 2019c). As a result of the measures aimed at reducing air pollution, the emission of pollutants decreased by 42% between 2006 and 2016 (Surgutneftegas 2016a, pp. 68 and 70). Surgutneftegas has consistently been ranked high in WWF Russia and CREON reports (2014, 2015, 2016) on the environmental responsibility of Russian oil companies. In 2014 and 2015, Surgutneftegas was rated first but dropped to third place in 2016 after Sakhalin Energy and Gazprom. In 2018, it was ranked number one in terms of environmental management.

COPING WITH CHANGE

Surgutneftegas is assumed to be one of the most conservative and least adaptable of the Russian oil majors. However, while many of its strategic principles have remained unchanged throughout its history, the company has more than once demonstrated its ability to respond to new challenges. In some cases, the response strategy has precisely involved sticking to the old ways of doing things.

Responding to Low Oil Prices, Crises and Sanctions

Throughout its history, Surgutneftegas has been able to cope successfully with financial and economic crises by adhering to its long-term strategy. Russia's economic meltdown in 1998 was an acid test for most Russian vertically integrated oil companies. Compared to its peers, the crisis had little impact on Surgutneftegas apart from lowering the value of its shares. The company's focus on self-reliance turned out to be a winning strategy in times of trouble. Since it did not have debts in foreign currencies, it has weathered both low oil prices and the dramatic depreciation of the rouble (NFGR 2006).

When the 2008 financial crisis began, the value of Surgutneftegas's short-term financial investments, cash and other current assets rose by 50% to RUB 383 billion. One manager explained that the company had accumulated financial reserves in the past 'in case of a possible crisis. You do not remember the times when there was no money to pay salaries' (Vedomosti 2008). Similarly, at the AGM in June 2009, Vladimir Bogdanov commented: 'We were always accused of channelling a lot of money to our investment reserve. Today we can see how justified this strategy was' (Vedomosti 2009b). Surgutneftegas was also the only vertically integrated oil company in Russia to recruit new people during the 2008 crisis. From mid-2008 to mid-2009, its headcount increased by 2% to 94 502 employees (although the company cut average salaries in 2009 compared to 2008) (Vedomosti 2009c).

Surgutneftegas recognizes oil price volatility as one of the key financial risks for the company. The *2016 Annual Report* notes that the company takes into account possible changes of oil and petroleum product prices using scenario planning for the development of investment projects and budgets, implements cost reduction programmes and re-evaluates the existing business plans (Surgutneftegas 2016a, p. 24).

Recognizing the impact of global oil price variations on its business, at the 2017 AGM, Bogdanov stated that 2016 had been unstable because of declining oil prices and currency fluctuations. Surgutneftegas then had some USD 34 billion in hard currency deposits, and because of the appreciation of the rouble, the company reported a loss of RUB 105 billion – its first in many years (Znak 2017). However, by October 2017, it had already increased its currency deposits to USD 37 billion.

Regarding Surgutneftegas savings kept on currency deposits (and again highlighting the risks associated with oil price volatility), Bogdanov said in 2013 that 'these funds are an insurance mechanism; nobody knows what will happen with the oil prices. We need this money so that the company can continue existing without problems. If the situation of 1998 is repeated, what would we do then?' (Kommersant 2013b).

Despite the losses that the company sustained in 2016, Surgutneftegas did not intend to revise its strategy. According to Bogdanov, 'it does not make sense to change financial policy trying to guess where the rouble rate of exchange or oil price would go. We are focused on our goals: ensuring production efficiency, cost reduction and introduction of technologies' (Znak 2016). However, Surgutneftegas is oriented towards an average oil price of USD 40–50 per barrel for the near future.

Interestingly, despite its investor-unfriendly reputation, according to Bloomberg, Surgutneftegas was the company with the highest dividend yield in the world in 2016. It was the only public oil company that ensured a positive return to investors after the November 2016 decision of OPEC and non-OPEC producers to reduce oil output. In 2015, the dividend yield of its shares was 18.5%, significantly higher than that of Shell (5.6%), for example. Analysts attributed these impressive results to its large dollar reserves. Also, since June 2014, when oil prices began to fall, Surgutneftegas was a leader in Russia both in a dividend yield of its preferred stock (37.8%) and total shareholder return (TSR) (96%) (Vedomosti 2016).

In September 2014, the United States introduced sanctions against several Russian oil companies, including Surgutneftegas. Restrictive measures forbade the US export of commodities, services and technologies to support deep-water, Arctic and shale projects of Gazprom, Gazprom Neft, LUKOIL, Rosneft and Surgutneftegas (Forbes 2014). The new US sanctions introduced in August 2017 applied to these five companies and forbade American citizens and companies from participating in their new projects both within and outside Russia (RBC 2017). However, Bogdanov commented that the sanctions would have little effect on Surgutneftegas since it did not have any debt denominated in foreign currencies, did not have any deep-water or (offshore) Arctic projects and had limited dependence on foreign technology (Neftegaz.ru 2014a).

Unconventional Reserves

Surgutneftegas is one of the few oil companies in Russia that can produce hard-to-recover reserves on its own without international partners. Since 2005, the company has been working on the Bazhenov shale oil formation in Western Siberia – the world's largest. By 2013, it had developed schemes for testing and commercial production in 19 oilfields and was operating 87 wells within these fields. But over the years, it produced only 2.5 million tonnes of oil, and the company insisted that the development of Bazhenov required state support (Surfutneftegas 2013).

In 2013, Surgutneftegas increased its oil production from the Bazhenov formation by 60% from the 2012 level. The company's chief geologist commented that although the company has incurred a 3 billion rouble loss, Surgutneftegas invests in the development because it is a promising direction for future operations (Neftegaz.ru 2014b). The company planned to commission 24–39 development wells annually. Surgutneftegas expected that by the end of 2018, the accumulated production from the Bazhenov formation would have reached 5.7 million tonnes (by 2014, it was about 2.5 million tonnes) (OGJ Russia 2014).

Climate Change

Since 1997, Surgutneftegas has had a policy of sustainable energy use and energy efficiency as a way to improve the company's financial performance (Surgutneftegas 2019a). The *2016 Annual Report* states that the 'rational use of energy resources, energy-saving and advanced engineering solutions in the power generation sector are among the main drivers to ensure better production performance and competitiveness of the company' (Surgutneftegas 2016a, p. 55). Thus, while the company and Bogdanov are not known for making statements about the importance of climate change, the company does pursue an efficiency-based policy which can help limit GHG emissions. Over a five-year period, the company reduced its sulphur dioxide emissions by 25% (Surgutneftgas 2019a). In the *2019 Annual Report*, Bogdanov claimed that the company directs significant funds to environmental programmes and environmental policy and that this is one of the key priorities of the company (Surgutneftgas 2019b).

Ninety-nine per cent of Surgutneftegas's natural gas is produced as APG – a by-product of oil extraction. Surgutneftegas makes a considerable contribution to climate change mitigation efforts in Russia by minimizing the flaring of APG in its fields. Surgutneftegas was the first Russian company to introduce gas utilization equipment and technology, an experience that is being emulated by its Russian peers. In the 1990s, it installed three gas-turbine power stations with a total capacity of 43.5 MW to utilize APG instead of flaring it, to meet the energy needs of the Tyanskoye and Konitlorskoye fields. As a result of this technology development, Surgutneftegas's level of APG utilization has been consistently the highest in Russia, that is, 99.56% in 2018, according to the annual report.

Surgutneftegas utilizes APG in the following main ways: processing at the Surgutneftegas gas-processing plant (some 62%) with subsequent deliveries to domestic consumers, power generation needs at the company's power plants (over 21%) and utilization as fuel and for technological needs of Surgutneftegas (15%) (Surgutneftegas 2016b, pp. 42–3). Surgutneftegas aims to further upgrade the system of APG gathering, transportation and utilization in the current fields and by establishing such systems in new fields.

As we have not been able to find any information from publicly available sources that Surgutneftegas or Vladimir Bogdanov personally regard climate change as an important issue, it seems ironic that Surgutneftegas simultaneously makes a considerable contribution to climate change mitigation efforts in Russia by successfully minimizing the flaring of APG from its fields. In 2016, Surgutneftegas started to implement a system of GHG emissions accounting to comply with Russian legislation and international standards (Surgutneftegas 2016b, p. 34). According to the *2016 Environmental Report*, Surgutneftegas reports on GHG emissions to the Russian Ministry of Energy, the Russian Federal State Statistics Service and other regulators (Surgutneftegas 2016b, p. 34).

CONCLUSION

Surgutneftegas is the most conservative oil company in Russia. Led by Vladimir Bogdanov for over 30 years, it has been stably successful during three distinct periods in modern Russian history: Soviet socialism, the gangster capitalism of the 1990s and the authoritarian capitalism of the early part of third millennium.

It has remained an independent company despite the efforts of other players to acquire it. Being consistently loyal to its long-term strategies and principles, Surgutneftegas has successfully adapted to new realities and found a winning formula for doing business in Russia. Its good ties with the political elite have helped it achieve its goals and be protected against hostile takeovers. However, at the same time, this protection might have cost Surgutneftegas its freedom to make strategic business decisions without the Kremlin's blessing. It has also had to serve a political agenda from time to time.

The company has remained resolutely non-transparent about its operations and shareholder structure. Uniquely among Russia's oil majors, Surgutneftegas has kept foreign partners and investors at arm's length. At the same time, it has made efforts to adopt and employ the West's best technologies. The company has always paid its taxes, avoiding tax scandals affecting other companies. It has addressed the climate change imperatives by adopting efficiency measures and reporting publicly on its emissions and environmental and social performance without exaggerating its achievements.

In the current era of climate change and the rise of alternatives to oil combined with the increasing international isolation of Russia, Surgutneftegas seems to be well placed among its peers to survive, having in place appropriate reporting systems, a culture of technology-based production efficiency and a long-standing strategy of independence from foreigners.

NOTE

1. Powerful actors with a background in the military or security services.

REFERENCES

BBC (2013), 'Pressa Britanii: V poiskakh vladeltsev Surgutneftegaza', accessed 15 February 2019 at https://www.bbc.com/russian/uk/2013/10/131024_brit_press.
Bloomberg (2015), 'Putin's next takeover target is oil giant's $34 billion cash pile', accessed 15 February 2019 at https://www.bloomberg.com/news/articles/2015-05-11/putin-s-34-billion-siberian-hoard-hunted-by-cash-starved-allies.
Burneft.ru (2017), 'O segodnyashnem dne krupneyshey neftyanoy kompanii Rossii OAO "Surgutneftegaz", vstrechayushchiy svoye 40-letiye', accessed 15 February 2019 at https://burneft.ru/archive/issues/2017-09/24.
Forbes (2001), 'The world's richest people', accessed 14 February 2019 at https://www.Forbes.com/2001/06/21/billionairesindex.html#3a9de4297e97.
Forbes (2004), 'Surgutskiy pasyans', accessed 15 February 2019 at http://www.Forbes.ru/Forbes/issue/2004-04/2365-surgutskii-pasyans.
Forbes (2014), 'SSHA vveli sanktsii protiv Lukoyla, Gazproma i Sberbanka', accessed 15 February 2019 at http://www.Forbes.ru/news/267601-ssha-vveli-novye-sanktsii-protiv-rossii.
Gazeta.ru (2013), 'Surgut nichem ne riskuyet', accessed 15 February 2019 at https://www.gazeta.ru/business/2013/04/30/5286761.shtml.
Greenpeace (2017), 'Reindeers in Moscow: Saving the sacred lake', accessed 15 February 2019 at https://www.greenpeace.org/archive-international/en/news/Blogs/makingwaves/reindeers-moscow-save-lake-numto-oil/blog/58887/.
Gustafson, T. (2012), *Wheel of Fortune: The Battle for Oil and Power in Russia*, Cambridge, MA: Belknap Press of Harvard University Press.
Ihodl.com (2015), 'Putin natselilsya na Surgutneftegaz', accessed 15 February 2019 at https://ru.ihodl.com/analytics/2015-05-13/putin-natselilsia-na-surgutneftegaz/.
Ino TV (2014), 'Die Welt: Surgutneftegaz-tayezhnaya tayna Putina', accessed 15 February 2019 at https://russian.rt.com/inotv/2014-07-08/Die-Welt-Surgutneftegaz---taezhnaya.
Kommersant (2000), 'Vse v nedrah. Krome lyudei i olenei', accessed 15 February 2019 at https://www.kommersant.ru/doc/16956.
Kommersant (2009), 'Natsionalniye media napolnilis Surgutneftegazom', accessed 15 February 2019 at https://www.kommersant.ru/daily/2009-03-04.
Kommersant (2011a), 'Surgutneftegaz ne puskayut v MOL', accessed 15 February 2019 at https://www.kommersant.ru/doc/1616250.
Kommersant (2011b), 'Surgutneftegaz otstupil bez poter', accessed 15 February 2019 at https://www.kommersant.ru/doc/1647101.
Kommersant (2012a), 'Surgutneftegas perelozhil aktivy', accessed 15 February 2019 at https://www.kommersant.ru/doc/2021024.
Kommersant (2012b), 'Neftyaniki opozdali s modernizatsiyei', accessed 15 February 2019 at https://www.kommersant.ru/doc/2002903.

Kommersant (2012c), 'Surgutneftegaz ostanovil padeniye dobychi nefti', accessed 15 February 2019 at https://www.kommersant.ru/doc/1853155.

Kommersant (2012d), 'Surgutneftegaz nachinayet tratit', accessed 15 February 2019 at https://www.kommersant.ru/doc/2093757.

Kommersant (2012e), 'Vladimir Bogdanov ne srabotalsya s venesuelskoi neftyu', accessed 15 February 2019 at https://www.kommersant.ru/doc/2030559.

Kommersant (2013a), 'Surgutneftegas otchitalsya standartno', accessed 15 February 2019 at https://www.kommersant.ru/doc/2183661.

Kommersant (2013b), 'Vladeltsi Surgutneftegaza sobralis v uzkom krugu', accessed 15 February 2019 at https://www.kommersant.ru/doc/2222921.

Kommersant (2017), 'Rossiya pristupayet k samym bogatym zalezham slantsevoy nefti v mire', accessed 15 February 2019 at https://www.kommersant.ru/doc/3329777.

Kompromat (2007), 'Novaya zagadka Surgutneftegaza', accessed 15 February 2019 at http://www.compromat.ru/page_19965.htm.

Kremlin (1992), 'Ukaz Prezidenta Rossiyskoy Federatsii ot 17.11.1992 g. No. 1403', accessed 15 February 2019 at http://www.kremlin.ru/acts/bank/2417.

Lenta.ru (2019), 'Sechin, Igor, Prezident NK Rosneft', accessed 15 February 2019 at https://lenta.ru/lib/14160890/full/.

Neft i kapital (2004), no title, p. 26.

Neft i kapital (2012), 'Novye tekhnologii povysyat koeffitsiyent izvlecheniya rossiys-koy nefti', accessed 15 February 2019 at https://oilcapital.ru/news/upstream/20-08-2012/novye-tehnologii-povysyat-koeffitsient-izvlecheniya-rossiyskoy-nefti.

Nefte.ru (2019), 'Surgutneftegaz', accessed 15 February 2019 at http://www.nefte.ru/company/rus/surgut.htm.

Neftegaz.ru (2008), 'Surgutneftegaz vstretilsya s korennymi zhitelyami Yugry', accessed 15 February 2019 at https://neftegaz.ru/news/view/85108-Surgutneftegaz-vstretilsya-s-korennymi-zhitelyami-Yugry.

Neftegaz.ru (2014a), 'Surgutneftegaz ne pochuvstvuyet sanktsiy blagodarya nalichiyu sobstvennykh tekhnologiy', accessed 18 February 2019 at https://neftegaz.ru/news/view/129751-Surgutneftegaz-ne-pochuvstvuet-sanktsiy-blagodarya-nalichiyu-sobstvennyh-tehnologiy.

Neftegaz.ru (2014b), 'Surgutneftegaz v 2013 g uvelichil dobychu nefti na mestorozh-deniyakh Bazhenovskoy svity na 60%', accessed 18 February 2019 at https://neftegaz.ru/news/view/119671-Surgutneftegaz-v-2013-g-uvelichil-dobychu-nefti-na-mestorozhdeniyah-Bazhenovskoy-svity-na-60.

Neftyaniki (2015), 'Zadacha-minimum: Spasti i sokhranit', accessed 15 February 2019 at http://www.nftn.ru/zadacha_minimum_spasti_i_sokhranit.

Newsruss.ru (2019), 'Surgutneftegaz', accessed 15 February 2019 at http://newsruss.ru/doc/index.php/Сургутнефтегаз.

NFGR (2006), 'Surgutneftegaz', accessed 15 February 2019 at http://www.ngfr.ru/library.html?surgutneftegas.

NIAR (2014), 'Surgutneftegaz planiruyet postroit truboprovod ot Kirishey do Ust-Lugi, ot proyekta v bukhte 'Batareynaya' kompaniya otkazalas', accessed 15 February 2019 at https://xn--c1aacfnblzfasadv.xn--p1ai/component/content/article/67-ust-luga/39440-surgutneftegaz-planiruet-postroit-truboprovod-ot-kirishej-do-ust-lugi-ot-proekta-v-bukhte-batarejnaya-kompaniya-otkazalas.

OGJ Russia (2014), 'Rossiyskiye kompanii vse bol'she investiruyut v slantsevyye proyekty', accessed 18 February 2019 at https://ogjrussia.com/uploads/images/Articles/July_14_article1/Resource2.pdf.

Overland, I., J. Godzimirski, L.P. Lunden and D. Fjaertoft (2013), 'Rosneft's offshore partnerships: The re-opening of the Russian petroleum frontier?', *Polar Record*, **49** (2), 140–53.

Poussenkova, N (2004), 'From rigs to riches: Oilmen vs. financiers in the Russian oil sector', accessed 14 February 2019 at http://www.bakerinstitute.org/research/from -rigs-to-riches-oilmen-vs-financiers-in-the-russian-oil-sector/.

RBC (2017), 'Novyye sanktsii SSHA prervut mezhdunarodnuyu ekspansiyu rossiys-kikh neftyanikov', accessed 18 February 2019 at https://www.rbc.ru/economics/04/ 08/2017/59824c529a7947b30d0818eb.

Ria Novosti (2004), 'Garvardskiy universitet podal v nyu-yorkskiy arbitrazh isk protiv "Surgutneftegaza"', accessed 15 February 2019 at https://ria.ru/economy/20040629/ 622568.html.

Russian Petroleum Investor (1997), *Russian Petroleum Investor*, December 1996– January 1997.

Shvarts, E.A., J. Bunina and A. Kniznikov (2015), 'Voluntary environmental stand-ards in key Russian industries: A comparative analysis', *International Journal of Sustainable Development and Planning*, **10** (3), 1–15.

Shvarts, E.A., A.M. Pakhalov and A.Y. Knizhnikov (2016), 'Assessment of environ-mental responsibility of oil and gas companies in Russia: The rating method, *Journal of Cleaner Production*, **127**, 143–51.

Shvarts, E., A. Pakhalov, A. Knizhnikov and L. Ametistova (2018), 'Environmental rating of oil and gas companies in Russia: How assessment affects environmental transparency and performance', *Business Strategy and the Environment*, **27** (7), 1023–38.

Surgutneftegas (2013), '469', accessed 4 April 2018 at http://www.surgutneftegas.ru/ press/smi/item/469.

Surgutneftegas (2016a), *Annual Report 2016*, accessed 15 February 2019 at https:// www.surgutneftegas.ru/en/investors/reporting/godovye-otchety/.

Surgutneftegas (2016b), *Environmental Report*, accessed 15 February 2019 at https:// www.surgutneftegas.ru/en/responsibility/ecology/ekologicheskie-otchety/.

Surgutneftegas (2018), *Annual Report 2018*, accessed 28 June 2019 at https://www .vestifinance.ru/articles/120606.

Surgutneftegas (2019a), 'Energoeffektivnost i resursosberezheniye', accessed 18 February 2019 at https://www.surgutneftegas.ru/responsibility/ecology/ prirodookhrannye-aspekty-khozyaystvennoy-deyatelnosti/energoeffektivnost-i -resursosberezhenie/.

Surgutneftegas (2019b), *Annual Report 2019*, accessed 18 February 2019 at https:// www.surgutneftegas.ru/en/investors/reporting/godovye-otchety/.

Surgutneftegas (2019c), 'Osnovniye napravlenya prirodochrannoi deyatelnosti', accessed 18 February 2019 at https://www.surgutneftegas.ru/responsibility/ecology/ prirodookhrannye-aspekty-khozyaystvennoy-deyatelnosti/osnovnye-napravleniya -prirodookhrannoy-deyatelnosti/.

Territory neftegas (2015), 'Territory neftegas', accessed 28 June 2019 at http://neftegas .info/upload/iblock/080/080093ba849aa9c64f5d3685e431badb.pdf.

The Moscow Times (1997), 'Obscure pension fund snaps up Surgut shares', accessed 15 February 2019 at http://old.themoscowtimes.com/sitemap/free/1997/2/article/ obscure-pension-fund-snaps-up-surgut-shares/311158.html.

Ugra.mk.ru (2012), 'OAO Surgutneftegaz: Sotsial'naya otvetstvennost kak prioritet', accessed 15 February 2019 at http://ugra.mk.ru/articles/2012/10/03/756094-oao -surgutneftegaz-sotsialnaya-otvetstvennost-kak-prioritet.html.

Vedomosti (2005a), 'Surgut popal na Ren-TV', accessed 15 February 2019 at https://www.vedomosti.ru/newspaper/articles/2005/09/02/surgut-popal-na-ren-tv.

Vedomosti (2005b), 'Surgut zaimetsya shelfom', accessed 15 February 2019 at https://www.vedomosti.ru/newspaper/articles/2005/12/05/surgut-zajmetsya-shelfom.

Vedomosti (2006), 'Voda i energiya dorozhe benzina', accessed 15 February 2019 at https://www.vedomosti.ru/newspaper/articles/2006/12/06/voda-i-jenergiya-dorozhe -benzina.

Vedomosti (2007a), 'Surgut ne fontaniruyet', accessed 15 February 2019 at https:// www.vedomosti.ru/newspaper/articles/2007/12/18/surgut-ne-fontaniruet.

Vedomosti (2007b), 'Pugayuschiye pensionery', accessed 15 February 2019 at https:// www.vedomosti.ru/newspaper/articles/2007/01/22/pugayuschie-pensionery.

Vedomosti (2008), 'Kompaniya nedeli: Surgutneftegaz', accessed 15 February 2019 at https://www.vedomosti.ru/newspaper/articles/2008/11/20/kompaniya-nedeli -surgutneftegaz.

Vedomosti (2009a), 'Ni vrag, ni drug', accessed 15 February 2019 at https://www .vedomosti.ru/newspaper/articles/2009/04/24/ni-vrag-ni-drug.

Vedomosti (2009b), 'Vybralis iz yamy', accessed 15 February 2019 at https://www .vedomosti.ru/newspaper/articles/2009/08/17/vybralis-iz-yamy.

Vedomosti (2009c), 'Shtatnaya ekonomiya', accessed 17 April 20019 at http://oil .rftoday.ru/tnk_bp/9/.

Vedomosti (2012a), 'Neftetreider Surguteks smenil vladeltsev', accessed 15 February 2019 at https://www.vedomosti.ru/business/articles/2012/07/26/troe_posle _timchenko.

Vedomosti (2012b), 'Chem zhivet Surgutneftegaz', accessed 15 February 2019 at https://www.vedomosti.ru/business/articles/2012/07/02/u_nih_neft_takaya_zhe.

Vedomosti (2013a), 'Surgutneftegaz ostaetsya v Rossii', accessed 15 February 2019 at https://www.vedomosti.ru/business/articles/2013/06/28/nachalos-godovoe-sobranie -akcionerov-surgutneftegaza.

Vedomosti (2013b), 'Surgutneftegaz sobirayet investorov', accessed 15 February 2019 at https://www.vedomosti.ru/business/articles/2013/04/03/surgutneftegaz_podarit _den_investoram.

Vedomosti (2013c), 'Surgutneftegaz ne raskryl vladeltsev kompanii', accessed 15 February 2019 at https://www.vedomosti.ru/business/articles/2013/05/06/surgut _bez_syurprizov.

Vedomosti (2014), 'Za kompaniei, kupivshei yuganksneftegas v 2004 godu, stoyal Surgutneftegaz', accessed 15 February 2019 at https://www.vedomosti.ru/business/ articles/2014/07/28/gaagskij-sud-za-kompaniej-bajkalfinansgrup-kupivshej.

Vedomosti (2016), 'Surgutneftegaz okazalsya samoy dokhodnoy neftyanoy kompani-yey v mire', accessed 15 February 2019 at https://www.vedomosti.ru/business/ articles/2016/02/20/630877-surgutneftegaz-samoi-dohodnoi.

Vestnik (2013), 'Chtob i neft dobyvalas, i oleni paslis', accessed 15 February 2019 at http://vestniksr.ru/news/6070-chtob-i-neft-dobyvalas-i-oleni-paslis.html.

WWF Russia (2014), 'WWF and Creon environmental responsibility rating of Russian oil and gas companies, 2014', accessed 14 February 2019 at http://wwf.ru/about/ what_we_do/oil/full_list/rating/eng.

WWF Russia (2015), 'WWF and Creon environmental responsibility rating of Russian oil and gas companies 2015', accessed 14 February 2019 at http://wwf.ru/about/ what_we_do/oil/full_list/rating/eng.

WWF Russia (2016), 'WWF and Creon environmental responsibility rating of Russian oil and gas companies 2016', accessed 14 February 2019 at http://wwf.ru/about/ what_we_do/oil/full_list/rating/eng.

Znak (2015), 'Zanachku Bogdanova raspechatayut dlya Sechina', accessed 15 February 2019 at https://www.znak.com/2015-05-12/zanachku_bogdanova_raspechatayut _dlya_sechina_surgutneftegaz_mogut_zastavit_vykupit_akcii_rosnefti_c.

Znak (2016), 'Bogdanov: Surgutneftegaz ne izmenit finansovuyu politiku, nesmotrya na ubytki v 2016 godu', accessed 15 February 2019 at https://www.znak.com/2017 -06-30/bogdanov_surgutneftegaz_ne_izmenit_finansovuyu_politiku_nesmotrya_na _ubytki_v_2016_godu.

Znak (2017), 'Bogdanov rasskazal aktsioneram Surgutneftegaza o situatsii v kompa- nii', accessed 15 February 2019 at https://www.znak.com/2017-06-29/bogdanov _rasskazal_akcioneram_surgutneftegaza_o_situacii_v_kompanii.

6. Tatneft: Genghis can

Tatneft, the oil company of the Republic of Tatarstan, is the fifth-largest Russian oil company in terms of crude oil production (after Rosneft, LUKOIL, Gazprom Neft and Surgutneftegas). It is also the only major Russian oil company with an ethnic identity. The Tatars are a distinctive Turkic-speaking, predominantly Muslim people who make up 53% of the population of Tatarstan. The republic was once seen as one of the more independently minded subjects of the Russian Federation, and this is reflected in the way Tatneft operates. Tatneft struggles valiantly with its depleting resource base and has been maintaining oil output levels by optimizing production, raising efficiency, promoting technological innovation and investing in extra-viscous crude. It is the only Russian oil company that has built a major new refinery since the collapse of the USSR; it is also one of the few that has a strong petrochemical arm. Tatneft has done well concerning import substitution and support for domestic producers as well as avoiding Western sanctions, mainly because the company was not thought to be as closely connected with the Kremlin as some other Russian oil companies. Tatneft has partnered with foreign companies within Russia; however, it has not been particularly successful in its attempts at investing abroad.

As of early 2019, the proved hydrocarbon reserves of Tatneft amounted to 971 million tonnes, while oil production in 2018 reached almost 30 million tonnes (Tatneft Press-Tsentr 2019). Tatneft accounts for over 80% of the oil produced in the Republic of Tatarstan and some 8% of Russia's crude output.

Tatneft is a fully vertically integrated company with refining, petrochemicals and gas-processing facilities. It has a network of 692 fuel stations, including 111 stations in Ukraine and 15 in Belarus. Tatneft is also a major producer of tyres, with three tyre-manufacturing plants based in the town of Nizhnekamsk in Tatarstan (Tatneft 2018a). Tatneft also has its own research institute, TatNIPIneft, located in Bugulma, Tatarstan.

Tatneft mainly operates in the Republic of Tatarstan, but it also has investments in the Samara, Orenburg and Ulyanovsk regions as well as the Republic of Kalmykia and Nenets Autonomous District. Tatneft's minimal presence abroad is in Belarus, Turkmenistan and Ukraine.

The General Director of Tatneft is Nayl Maganov. Like most other Russian oil and gas companies, except LUKOIL and Surgutneftegas, Tatneft's leadership has changed numerous times since the Soviet era. Maganov has been the company's general director since November 2013, taking over from Shafagat Takhautdinov. Maganov previously served in senior management positions in Tatneft when the company was privatized in 1994. Prior to that, he had worked for NGDU Elkhovneft, part of the Tatneft Production Association.

As of May 2015, the company's key shareholders were the National Settlement Depository (59.55%) and OJSC Central Depository of Republic of Tatarstan (30.45%). Tatneft's depository receipts are listed on the London Stock Exchange and are traded in the Xetra system within Deutsche Börse (Tatneft 2018b).

CORPORATE HISTORY

Commercial oil production in Tatarstan commenced in 1943, making it one of Russia's oldest petroleum provinces. The super-giant Romashkinskoye field is among the top ten oilfields in world history and was discovered in 1948, transforming Tatarstan into the most oil-rich region in the USSR. At the time, Tatarstan's reserves amounted to 2.3 billion tonnes, including recoverable reserves of 1.4 billion tonnes. Most of Tatarstan's crude is of low quality, with medium to high viscosity and an average sulphur content of 2.5%.

Tatneft's predecessor, the Tatneft Production Association, was established in January 1949. In 1956, Tatneft extracted 18 million tonnes of oil, becoming the number one crude producer in the Soviet Union. Tatarstan demonstrated high petroleum output growth rates with the volume of oil production rising from 43 million tonnes in 1960 to 104 million tonnes in 1975. By 1981, the accumulated hydrocarbon production in Tatarstan had reached 2 billion tonnes of oil and over 70 billion cubic metres of gas. However, 1975 was the last year that witnessed a rise in crude production. It subsequently began to decline steadily, mainly because of ageing fields, and reached a low of 24 million tonnes in 1994.

Because of its complex reserve base, Tatneft has always focused on innovation and new technology. For example, as early as 1984 and 1985, Tatneft introduced improved reservoir management techniques, which helped arrest the rapidly growing proportion of water in the oil it extracted (Lazard Capital Markets 1996, p. 13). During the 1980s, Tatneft's employees worked on several fields in Western Siberia on a fly-in, fly-out basis (Tatneft 2018c).

The 1990s

When market reforms began in Russia, the Republic of Tatarstan was in a different position from other Russian regions because the Kremlin was particularly keen to secure the loyalty of this ethnic Turkic, Muslim republic located in the centre of the country. To this end, an agreement between the federal authorities and Tatarstan on the division of competencies and mutual delegation of authority was signed in February 1994, giving Tatarstan greater privileges than most parts of the Russian Federation. The petroleum sector was regulated by special agreements. Therefore, Tatarstan was able to pursue a relatively independent energy policy and ensure significant fiscal benefits for Tatneft to encourage the company to continue production from wells with low daily output and/or a high water-cut, commissioning of new fields and introduction of new methods of enhanced oil recovery.

In 1994, Tatneft was privatized. Its authorized capital was split into ordinary shares (93.66%) and preference shares (6.34%). In October 1994, 2.3% of the authorized capital was sold at auctions. In accordance with the privatization plan, 30% of the authorized capital was sold at a discount or transferred free of charge to employees. Preference shares were also given to personnel, including 5.16% sold for privatization vouchers. Another large shareholder was the State Property Committee of Tatarstan, which held 40% plus the 'golden share'. Ten per cent of shares were given to investment funds. Other smaller stakes were sold to top managers, suppliers and residents of the oil-producing regions of Tatarstan (Khisamov 2007).

Owing to the favourable tax regime, Tatneft managed to stabilize crude production at 24 million tonnes per annum beginning in 1995. A decree signed by Tatarstan's then President Mintimer Shaimiev in 1997 created some 30 small oil companies operating in Tatarstan in addition to Tatneft, which provided diversification and competition for the larger company (Khisamov 2007).

During the 1990s, Tatneft was a Russian pioneer in the stock markets. Its shares were first traded over the counter in April 1995, and in September 1995 they were listed on the Russian Trading System (RTS). In June 1996, about 0.1% of its shares were traded in American depository receipt (ADR) form on NASDAQ's bulletin board, with the Bank of New York acting as depository.

By 1996, Tatneft had provided two years of US GAAP accounts, and Miller and Lents undertook an independent reserve audit. In light of this, investment analysts noted that no other Russian oil company had issued similar quality information (Lazard Capital Markets 1996, p. 4). In December 1996, Tatneft's ADRs were listed on the London Stock Exchange (LSE) and two years later on the New York Stock Exchange (NYSE), where 11.5 % of its shares were traded in the form of ADRs. At that point, Tatneft was the only oil company in Russia that had listings on two major international exchanges.

In December 1996, Tatarstan offered part of its stake in Tatneft for sale to global investors. After that, the government's share shrank to some 30%, but it was to retain the 'golden share' until 1999 (Lazard Capital Markets 1996, p. 1). Russian experts regarded the December 1996 sale of Tatneft's shares as the most significant stock market event related to Russian oil companies that year. Even a statement by the Federal Agency on Insolvency about the alleged imminent bankruptcy of Tatneft made on the eve of the private presentation that the company gave to Western investors did not mar the event (Savushkin 1997). It subsequently turned out that Tatneft did not go bankrupt: it reached an agreement with the federal authorities, probably with the help of Tatarstan's political leaders, who did their best to protect the republic's enterprises (Melnikov 1996).

Tatneft entered the debt market in 1997 when it issued Eurobonds worth USD 300 million (Mazneva 2010). In 1998, Tatneft became the first Russian company to undergo the procedure of official listing of common and preferred stocks in the RTS. In 1999, it received listing for ordinary and preference shares at the Moscow Interbank Currency Exchange.

During the 1990s, Tatneft actively searched for international partners to help it overcome the inevitable problems associated with its declining reserve base, although its in-house engineers were quite advanced and technologically sophisticated and TatNIPIneft delivered high-class research results and registered valuable innovations. By the mid-1990s, Tatneft had established three joint ventures with foreigners. The most important of them was Tatalpetro with Total (Tatneft owned 50%). Its objective was to increase oil production in the Romashkinskoye field through the use of modern enhanced oil recovery (EOR) techniques, including chemical injection. Tatex, a joint venture with the American company Global Natural Resources (Tatneft had 50%), installed Western equipment to increase the recovery of liquids from associated petroleum gas (APG) streams. Tatoilgaz was Tatneft's first production-oriented joint venture, with the German company Mineralol Rohstoff Handel (MRH). This joint venture was engaged primarily in reviving old fields with a combination of horizontal drilling, hydrofracking and secondary recovery. An associate of MRH also cooperated with Tatneft in setting up fuel stations in Tatarstan (Lazard Capital Markets 1996, p. 15).

Initially, Tatneft was exclusively established as an exploration and production (E&P) company with insignificant refining capabilities equivalent to less than 2% of its oil output. However, by the late 1990s, Tatneft had acquired major stakes in petrochemical enterprises in Tatarstan and had actively started building its network of fuel stations, thus gradually transforming itself into a vertically integrated company.

During the mid-1990s, Tatneft tried to secure stable markets for its heavy and sour crude by attempting to establish three financial-industrial groups: Mostatnafta (based on a refinery in Moscow), Volga-Oil (based on a refinery in Nizhniy Novgorod) and Ukrtatnafta (based on a refinery in Kremenchug). However, all three attempts failed for various reasons, and in the following decade, Tatneft was forced to build its own refinery for processing its oil.

The 2000s

Changes in President Vladimir Putin's Russia affected Tatneft, particularly in the fiscal sphere. From the beginning of 2002, the profit tax rate was reduced in Russia and a flat mineral production tax was introduced in place of several previous charges for the use of subsurface resources. YUKOS and Sibneft, with their high-quality reserves and thus low lifting costs, were the main beneficiaries of this initiative (which they had strongly lobbied), while Tatneft (and its neighbour Bashneft), with their inferior reserves and mature fields depleted by some 80% resulting in high lifting costs, suffered from this change. Also, Moscow began to feel uncomfortable about the independence of the regions and started to suppress them and especially Tatarstan, which was perceived to be the most independent of all. For example, in 2001, Tatarstan lost the privileges granted to it in 1994, and Tatneft was deprived of the fiscal benefits that it had enjoyed (Poussenkova 2010).

In the new millennium, Tatneft continued to be plagued by its declining reserve base, and in 2006, it became the first oil company in Russia to announce the downgrading of its proved reserves. A repeat audit conducted by Miller and Lents revealed that Tatneft's proved reserves of oil as of 1 January 2005 amounted to 5801 billion barrels, whereas it had earlier reported 5965 billion barrels. Following this announcement, Tatneft's market capitalization fell by USD 350 million to USD 12 billion (Tutushkin and Borisov 2006). Tatneft sought to combat the combined challenges of the declining reserve base and the new fiscal rules through an adaptive management and production strategy, including the expansion of its refining capabilities with the construction of its TANECO refinery (see the section on production strategy below).

Although Tatneft mainly focused on production issues and never played high-profile power games (in contrast to many other Russian oil companies), during the 2000s, as a minority shareholder, it became indirectly involved in one of the most notorious corporate conflicts of the decade around the refinery that was located in Moscow. In 1997, President Boris Yeltsin had issued a decree whereby Moscow was to receive 51% of voting shares in the Moscow refinery. In 1999, the Moscow city authorities united the refinery with Sibir Energy to form a fully vertically integrated company – the Moscow Oil and Gas Company, 69% of which was owned by Moscow and 31% by Sibir Energy (Khrennikov and Shevelkova 2007). The struggle for the control of the Moscow refinery between its main shareholders – Sibneft and Tatneft – and Moscow Oil and Gas Company lasted six years (see the chapter on Gazprom Neft in this book). However, when Gazprom bought 72% of Sibneft, it sought to resolve the issue. In 2007, peaceful negotiations began, and in early 2008, the shareholders agreed to manage the enterprise jointly, and a memorandum was signed between Yuri Luzhkov, the then Mayor of Moscow, and Alexei Miller, the CEO of Gazprom (Tutushkin 2008).

In the first two decades of the twenty-first century, Tatneft's stock market performance was somewhat patchy in contrast to the previous decade when it was described as a trailblazer in terms of information disclosure and corporate governance. This was reflected in the increasingly divergent perception of its performance by the key international rating agencies, which had been fairly unanimous in their positive assessments of Tatneft during the previous decade. Thus, in 2003, Fitch upgraded its rating of Tatneft's obligations from B- to B. Yet Tatneft was consistently late in providing its financial statements in the mid-2000s. In late 2004, 'due to limited transparency and disclosure as well as aggressive acquisition plans', S&P downgraded its rating from B to B- and, in August 2006, even revoked the B- rating following delayed financial statements. By contrast, in the same year, Fitch upgraded Tatneft's forecast rating to 'positive' because it expected that Tatneft could significantly improve its business structure following the launch of its new refinery (Mazneva and Malkova 2006). Moreover, in June 2006, Tatneft announced that it intended to delist its ADRs from the NYSE, and the exchange stopped trading Tatneft's securities in September 2006 (Panov and Gubeydullina 2006).

In the mid- to late 2000s, Tatneft explored opportunities to expand its upstream and downstream operations overseas with limited success (see the section on internationalization below). On the other hand, Tatneft benefited from some domestic policy developments. In July 2006, a federal law was adopted on the differentiation of mineral production tax for fields depleted by more than 80%, and a zero mineral production tax for super-viscous oil was introduced. These benefits were applicable to the Romashkinskoye field that had been depleted by 84% by that time, as well as four other Tatneft fields (Takhautdinov 2007).

Thus, despite a few setbacks, Tatneft remained among the best performing Russian stocks between 2005 and 2012 since its share price rose five-fold over the period, as observed by Sberbank analysts. The analysts attributed this performance to the company's fundamental changes. First, it allayed concerns that its production was facing an imminent decline. Second, it managed to de-leverage while maintaining a net cash surplus for three years running before the construction of the TANECO refinery began. Tatneft was not involved in any major mergers and acquisitions (M&A) or other projects. In addition, Tatneft improved disclosure: it began publishing Management Discussion and Analysis (MD&A) in 2006 and moved to quarterly reporting and regular calls with investors in 2008. It also increased dividends to 30% in 2006 and, in 2007, secured mineral production tax benefits on its depleted fields (Sberbank CIB 2014).

Post-2010

In recent years, Tatneft's position in global credit ratings has significantly improved. In November 2017, Fitch confirmed Tatneft's credit rating at BBB-with a 'stable' forecast. This reflected Tatneft's strong financial position following the commissioning and further development of TANECO. Compared to its Russian peers, the company also had a meagre ratio of debt to cash flow from operations (Tatneft Press-Tsentr 2017a). In January 2018, Moody's upgraded Tatneft's credit rating to Baa3 with a 'positive' forecast (Tatneft Press-Tsentr 2017b).

In 2016, together with LUKOIL and Novatek, Tatneft was included (as number five) in the top ten companies of the global petroleum sector by the Boston Consulting Group (BCG) (Tatneft Press-Tsentr 2017c). During that period, Russian experts also agreed that Tatneft, LUKOIL and Novatek were among the most profitable companies in Russia in terms of shareholder returns (Makhnyeva 2016). According to these experts, Tatneft was well known for its efficient management of existing projects, good communication with investors and low debt burden compared to its peers (Makhnyeva 2016). In 2017, Tatneft retained its top ten position in the BCG rating with a slight drop to number eight. As BCG noted, Tatneft kept a strong position in the rating due to higher operating efficiency, a low level of debt and a balanced dividend policy (Tatneft Press-Tsentr 2017c).

In 2016, Tatneft also became number one on the list of the most innovative European oil and gas exploration companies, according to the Thomson Reuters report *The Future is Open: 2016 State of Innovation* (cited in Tatneft Press-Tsentr 2016a). Tatneft may deserve this accolade, as it registered its five-thousandth patent for inventions in 2012 (Tatneft 2018d).

COMPANY PROFILE

Production Strategy

Tatneft currently produces the lion's share of its oil from six major fields (Romashkinskoye, Bavlinskoye, Sabanchinskoye, Novo-Elkhovskoye, Pervomaiskoye and Bondyuzhskoye), with Romashkinskoye accounting for the bulk of crude output: 15.8 million tonnes in 2016 (out of 28.3 million tonnes produced in total by Tatneft) (Tatneft 2018e). Tatneft delivers oil to the domestic market (15 million tonnes per annum), the former Soviet Union (1.3 million tonnes per annum) and other overseas destinations (10.4 million tonnes per annum). Domestic deliveries are mainly made to the TANECO refinery (8.6 million tonnes) and TAIF-NK (6.2 million tonnes) (Tatneft 2018f).

TAIF-NK is a major refining complex located in Nizhnekamsk, Tatarstan. It includes a refinery, gasoline plant and a facility for processing gas condensate.

Tatneft's production strategy is determined by the fundamental problems and weaknesses of its reserve base in Tatarstan, an ageing petroleum province. Tatneft faces a serious challenge from depleting fields and a low average daily output of operating wells – 3.9 tonnes per day with an average output of new wells of only 10 tonnes per day (Tatneft 2017). Having to cope with the problems presented by the continuously ageing reserve base as well as the tougher fiscal conditions imposed on its operations, Tatneft has been forced to adapt to these changes and adjust its strategies accordingly.

First, Tatneft sought to optimize its management structure and expand its activities beyond E&P. In 2002, it added gas and petrochemical branches to the business (Tatneftegazpererabotka and Tatneft-Neftekhim, respectively). In 2008, the company undertook large-scale restructuring and divested itself of some non-core assets, reorganized its system of managing the oil services sector and formed managing companies for different activities. In 2008, probably as a result of this streamlining, Tatneft was included for the first time in the Platts global rating of the most efficient energy companies (Top250). In 2010, Tatneft launched another new line of business, heat and power generation.

Second, Tatneft searched for innovative ways to increase its resources and maintain oil production levels. It began to develop Tatarstan's significant resources of super-viscous (or bituminous) oil, which requires heating to be extracted. In 2003, Tatneft started exploration and test production in the Ashalchinsk field of super-viscous oil. In 2006, Tatneft drilled a unique horizontal well there and produced its first 100 tonnes of bituminous crude (Tatneft 2018d). It also started expanding into new regions. It entered the Timan-Pechora province in the mid-2000s through two companies, Severgaznefteprom and Severgeologiya, which owned prospecting and E&P licences for four blocks in Nenets Autonomous District (Tatneft had a 50% stake in both companies) (Tatneft 2018d).

Third, Tatneft continued its transformation into a fully vertically integrated oil company by establishing its own refining facilities. In fact, it became the only Russian vertically integrated oil company that built a major refinery from scratch after the collapse of the USSR. Before that, it had only minor refining facilities at NGDU Elhovneft, part of the old Tatneft Production Association. In summer 2005, the leadership of Tatarstan decided to build an oil-refining complex in the republic. The project operator was to be an especially established closed JSC Nizhnekamsk refinery founded by Tatneft (40%), the state holding Svyazinvestneftekhim (9%) and two offshore companies that held a blocking stake. The plan was to build a 7-million-tonne-per-annum refinery (to be expanded later to 14 million tonnes per annum) – to be named TANECO – and a petrochemical facility in Nizhnekamsk.

Tatneft was to take care of financing, and it found an unusual way to raise funds. Its subsidiary Tatneft Oil AG became the co-founder of an investment fund, the International Petro-Chemical Growth Fund (registered on the island of Jersey). Tatneft contributed 5% of its ordinary shares to the authorized capital of the fund, the value of which was estimated at that time to be USD 485 million (Tatneft 2018d). In early 2006, Tatneft decided not to wait for external investors to finance the construction of the new TANECO refinery and, instead, allocated over USD 200 million of its own funds to the project. Experts noted that since the company was not actively buying new assets at the time, it had enough cash to fund the project (Borisov 2006). In 2006, TANECO also received federal support: The Russian Federation Investment Fund financed the design and construction of the external transportation infrastructure, including an oil pipeline, petroleum product pipeline and railroad facilities. In 2011, the first phase of the TANECO complex was commissioned – a facility for primary oil refining with a capacity for 7 million tonnes per annum. Tatneft and Chevron Lummus Global have been cooperating on projects related to the refinery since 2006 when they signed a licensing agreement and an agreement to design hydrocracking installations for TANECO (see below).

In 2016, Tatneft formulated its Strategy 2025, which envisaged the transition from stabilization to sustainable organic growth in the E&P segment with crude output to surpass 30 million tonnes by 2025. The company also aimed at reducing operating costs by 10% (Tatneft Strategy 2016). According to Strategy 2025, Tatneft's exploration division aimed for 100% replacement of its economically recoverable resource base through the following measures:

1. Increased efficiency in conventional oil exploration within mature fields.
2. Study and follow-up exploration of missed plays.
3. Exploration of new super-viscous oilfields.
4. Exploration of new fields in Nenets Autonomous District, Samara, Orenburg, Ulyanovsk regions and Kalmykia.
5. Exploration of oil in the Domanic formation in Tatarstan (Tatneft 2016).

Besides traditional seismic exploration, Tatneft is using several new technologies, such as artificial intelligence, geochemical exploration for oil and gas layers with the use of passive adsorption of hydrocarbons, low-frequency seismic sondage and electromagnetic sondage (Tatneft 2018g).

In the production sphere, the company continues to be strongly focused on super-viscous oil. Between 2017 and 2018, Tatneft planned to invest around RUB 20 billion into developing its super-viscous fields (20% of the overall investment programme of RUB 100 billion for this period). High-viscosity oil resources in Tatarstan are estimated to exceed 1.4 billion tonnes, and Tatneft hopes that it will be able to substitute the ageing fields with this type of crude in the future (Astakhova 2017; Tatneft Press-Tsentr 2016b).

Tatneft produced 1.9 million tonnes of super-viscous oil in 2018 (Finam 2019). The company considers bituminous crude to be a priority for growing its resource base. It is currently developing seven layers of super-viscous oil: four layers in Ashalchinsk field and one each in Languevsk, Karmalinsk and Nizhne-Karmalsk.

Nayl Maganov, Tatneft's General Director, admits that super-viscous oil is a complicated, knowledge-intensive and costly business, but still sees it as promising (Zavalishina 2015). It requires major investments that it is hoped will bring results in the form of produced oil in the future. Bituminous oil might become a crucial reserve for Russia's petroleum industry in years to come because resources of conventional crude in Russia are being depleted. According to Maganov, the mining and geological conditions of bitumen fields in Tatarstan are much more challenging than in Canada: bitumen is produced in the Canadian tundra while Tatneft's fields are located in populated areas of Tatarstan. Therefore, Tatneft must channel considerable funds into putting the territories in order, recultivating land and imposing strict environmental monitoring (Zavalishina 2015).

Unlike many of its Russian peers, Tatneft continues to focus much of its long-term production strategy on its refining capacities, and it is constantly developing its TANECO refinery. In December 2014, using a Chevron licence, TANECO commissioned a combined installation for hydrocracking vacuum gasoil. This permitted TANECO to launch the production of Euro-5 diesel, aviation kerosene and basic oils. The Euro-5 diesel fuel produced by TANECO is relatively clean with a sulphur content of less than 3 parts per million (ppm), while the threshold indicator for Euro-5 is 10 ppm.

In early 2016, Tatneft bought an additional 9% of TANECO and obtained full control over the enterprise. According to Tatneft's Strategy 2025, by that year the refinery's installed capacity should reach 14 million tonnes per annum; the conversion ratio will be 97% (compared to 74% in 2014) and the yield of light products 90% (compared to 69% in 2014). These would be high levels by current Russian standards (Neft i kapital 2014a; Tatneft Strategy 2016). In October 2017, TANECO and Chevron Lummus Global signed a MoU to enhance the efficiency of TANECO (Tatneft Press-Tsentr 2017d). Tatneft consistently supports domestic producers by placing orders for equipment in Russian enterprises; for instance, the Izhora Machine-Building plant has manufactured hydrocracking reactors, a product that was unique for Russia (Tatneft Press-Tsentr 2018).

Unlike many of its Russian peers, Tatneft has a well-developed petrochemical arm, comprising three tyre-manufacturing plants and several support enterprises, including hi-tech services for the petrochemical and tyre industries and Russia's largest producer of technical carbon (a key component of rubber).

Prompted by the complexity of its reserve base, Tatneft consistently pursues a policy of innovation, applying the most advanced technologies throughout the company. Its research institute TatNIPIneft has always been one of the most respected R&D establishments in the former Soviet Union and is responsible for several regionally significant advances in production technology and reservoir engineering. In 2014, Tatneft began the digital transformation of its whole production system, and by 2016, practically all branches of the company had been transferred to new information platforms. Currently, Tatneft is working on artificial intelligence and creating digital 'twins' of real objects. IT is not only used to revitalize Tatneft's own operations, but also developed and marketed for other companies.

Internationalization

Tatneft has a minimal presence abroad. Upstream, it now mainly operates in Turkmenistan. In 2008, Tatneft and Turkmenneft signed a protocol on the development of oil and gas cooperation in Turkmenistan, and in 2010, the two companies entered into a production contract. Tatneft subsequently opened a subsidiary in Turkmenistan to provide enhanced oil recovery services and maximize the potential of the Goturdepe field. In 2012, Tatneft began well operations under a service contract between Tatneft and Turkmennebit (Tatneft 2018h).

In the early 2000s, Tatneft signed a service contract with the Iraqi government; however, it had to leave urgently following the US invasion of 2003. In 2009, Tatneft qualified to participate in the second Iraqi licensing round but failed to win (Mazneva and Tutushkin 2009). In 2005, Tatneft was the first Russian company to enter Libya, winning the largest share of a 2006 licensing round and securing a PSA to work on four blocks with the National Oil Corporation of Libya. In 2009, Tatneft achieved its first commercial oil when testing an exploration well. However, the political unrest in Libya forced Tatneft to evacuate hundreds of workers and suspend its businesses in the country (Soldatkin 2011). In 2010, Tatneft began operating in Syria; however, it had to halt its operations because of escalating violence. In October 2016, Tatneft signed a MoU with the NIOC. Tatneft planned to study opportunities to develop Iran's Dehloran field (Tatneft Press-Tsentr 2016b). In early January 2018, Nayl Maganov announced that Tatneft had presented to NIOC its proposals for the development of Dehloran and Shadegan fields (Finance-Rambler 2018); however, no real progress has been reported since then.

Tatneft has been even less successful in developing its downstream activities overseas. For instance, in early 2004, a consortium of the Turkish Zorlu

Holding and Tatneft won 65.7% of the Turkish oil-refining holding TUPRAS (total capacity 26.6 million tonnes per annum). However, the deal was halted by the local trade unions and minority shareholders of TUPRAS, who initiated numerous lawsuits (Neft i kapital 2004).

Tatneft's experience in Ukraine has been particularly unsettling. In autumn 2006, Tatarstan transferred to Tatneft its stake in Ukrtatnafta that controlled the Kremenchug refinery in Ukraine (Tutushkin 2006). However, a corporate raid in October 2007 by the Ukrainian special forces caused the leadership of Ukrtatnafta to be replaced by the former chairman of its managing board, Pavel Ovcharenko. This followed a protracted conflict between the Ukrainian authorities and Naftogaz Ukraine, on the one hand, and the refinery management, controlled by Tatarstan, on the other. Prior to the raid, the Supreme Court of Ukraine had made a decision to transfer a controlling interest in the joint venture to Naftogaz. Tatneft disputed this in court and, following the raid, halted deliveries of its crude to Kremenchug (Tutushkin et al. 2007). It also bought two shareholders of Ukrtatnafta to strengthen its position in court (Mazneva 2008).

In 2008, Tatneft filed a lawsuit in Zurich against Ukraine, demanding the reimbursement of USD 1.1 billion for damages (Malkova and Kazmin 2008). Since 2008, the claim has risen to USD 2.4 billion. In 2014, the International Arbitration Court in The Hague reached a verdict in favour of Tatneft, but only for USD 112 million, and Ukraine filed a counter-suit in the Court of Appeal in Paris (Kozlov 2016). In early 2016, Tatneft filed a lawsuit in the British courts to recover USD 334 million of debt with interest from Ukrtatnafta for deliveries of oil to Kremenchug. In March 2016, Tatneft began court proceedings in the UK against Pavel Ovcharenko and three Ukrainian oligarchs concerning non-payment for oil delivered to the refinery in 2007. In November 2016, the Supreme Court of London declined the lawsuit on the grounds that it had no prospects of success. However, on 18 October 2017, the Court of Appeal of England and Wales revoked this decision (Tatneft Press-Tsentr 2017e). The legal war between Tatneft and Ukraine continues, having been described by *Kommersant* as one of the most important corporate conflicts of 2017 (Zanina and Rayskiy 2017). Also, in 2017, Tatneft filed a lawsuit against Ukraine in the Arbitration Court of Moscow, and this litigation continued in 2018 (Interfax 2018).

Shale Oil

Like its Russian peers, Tatneft was late to take an interest in shale oil due to its failure to foresee its importance. This is particularly interesting in the case of Tatneft as it specializes in extracting other difficult oil, notably extra-viscous oil.

Still, in response to the question of whether Russia and the rest of the oil-producing countries slept through the shale revolution, General Director Nayl Maganov made the following observation in an interview with Kommersant-FM:

> The question is, what do we mean by 'slept through'? Oil in shale rocks was known for a long time; since the 1960s–70s, people were quietly dealing with it; quietly trying to produce this oil. These technologies have continuously been developing since the 1960s, and at some point, they reached a level at which production became profitable. I do not exactly understand what revolution and who slept through it. (Kuzichev and Bogdanov 2016)

Tatneft began studying shale oil prospects more intensively in 2013 by drilling a pilot well to a depth of some 1700 metres. Tatneft's top management fully recognized the implications of the shale revolution in its 2012 Annual Report: 'The recent changes in the global oil and gas sector in which shale hydrocarbons began to play an increasingly prominent role could not help but make an impact on the major trends and priorities for the development of the Russian oil industry' (Gariffulin et al. 2012).

A report on the exploration and implementation of pilot shale oil projects was presented at a regular meeting of Tatneft's board of directors chaired by the republic's President Minnikhanov in Kazan in 2014. It stated that hydrofracking had been carried out in the Dankov-Lebedyanskiy reservoirs of the Bavlinskoye oilfield. The board of directors approved the shale programme for 2014–15; it also approved the establishment of testing grounds for the Domanic shale oil in Tatarstan (Oil & Gas Eurasia 2014).

In 2015, Tatneft booked some 29 million tonnes of shale oil reserves. In total, it planned to invest about RUB 1.5 billion under its 2016–18 shale programme. Tatneft also considered opportunities to join shale projects in the West. As Nayl Maganov admitted in an interview with *Business Gazeta*, the company's experts were invited by their American colleagues to visit the United States and study the prospects for Tatneft (Zavalishina 2015). In its 2016 Annual Report, Tatneft reconfirmed that the 'implementation of the pilot programme for shale oil production was among the company's priorities' (Voskoboinikov et al. 2016).

CSR

Since 2005, Tatneft has been publishing annual reports on sustainable development and social responsibility aimed at business partners, public authorities, employees, communities and civil society. The introduction to the 2013 report was as follows:

> In 2013, Tatneft company continued implementing principles of social responsibil- ity in the company's business practices. During the reporting period, considerable attention was traditionally paid to the environment, the creation of safe working conditions, social protection and the professional development of employees. A sig- nificant contribution was made to support healthcare, education, culture, sports and development of social infrastructure in the regions of activity.

Tatneft mainly operates in the central part of European Russia, a region with a relatively mild climate and a well-developed infrastructure. In contrast to the oil companies working in Western Siberia, Tatneft has not had to build oil towns from the ground up nor does it come into close contact with any Arctic indigenous peoples. On the other hand, it has greater visibility because it operates in more densely populated areas, and any environmental accident is likely to draw immediate large-scale attention. Moreover, its strong connection to the Tatarstan government and association with the Tatars as an ethnic group ensure that it is still under pressure to provide social services like other Russian oil companies.

Tatneft's CSR policy has several components. Much of its effort is focused on social projects. It also has projects related to environmental protection and the promotion of climate-friendly technology (see below). This comes in addi- tion to its efforts to increase efficiency and reduce GHG emissions in its core operations (see the section on climate change below).

In collaboration with the local municipalities, Tatneft implements projects to improve life in the towns and settlements in the regions where it works. As part of the programme aimed at supplying Tatarstan's population with clean drinking water, a water-pumping tower was built in Molodezhniy, Almetievsk region, and a water pipeline grid was installed in the village of Mencha (Ilmukova et al. 2013, p. 80). General Director Nayl Maganov considers support for medical services to be the most important type of social activity for the company (Zavalishina 2015). Projects have included re-equipping a new oncology centre, modernizing a children's polyclinic in Almetievsk as well as supporting the construction of an emergency ward, an obstetrics hospital, a regional medical diagnostic centre and inter-regional oncological polyclinics in various cities of Tatarstan (Ilmukova et al. 2013, p. 83; Tatar-Inform 2017).

Tatneft occasionally finances the construction and reconstruction of mosques, cathedrals and churches; it has built kindergartens and, in one case, an ice palace. In 2004, Tatneft established a 'talented children' fund to support scientific and creative projects and participation in conferences (Ilmukova et al. 2013, p. 82). It finances the Ruhiyat Fund of Spiritual Renaissance (founded in 1997), which supports children's creative development, festivals, publications and art and literature contests; it also established a literary scholarship prize in the name of Sazhida Suleimanova, a poetess. The charitable fund Mercy was founded to provide targeted support for orphans, veterans, the disabled and low-income families (Tatneft 2018j). Much of this charitable activity is typical of Soviet and Russian companies. However, in a move *not* typical of a Russian oil company, in 2017, Tatneft launched a social project to lease bicycles to the residents of Almetievsk at a low cost (Krivopatre 2017).

Regarding its environmental commitments, Tatneft has demonstrated leadership in developing and installing environmentally friendly technology and has made efforts to reduce its climate impact and enhance the efficiency of its operations (see the section on climate change below). Some of the company's environmental programmes have been designed along traditional Soviet-era lines, such as its programmes to plant trees along highways and oilfield routes. However, other efforts have indicated a commitment to stimulating good environmental performance in the energy sector within the company and beyond through technology development and promotion of new technologies. For instance, in January 2018, TANECO became the first enterprise in Russia to introduce automatic monitoring of emissions into the atmosphere (Tatneft Press-Tsentr 2018). Tatneft also seeks ways to enhance the efficiency of its petrochemical arm. Thus, in November 2014, Tatneft and the Italian company Marangoni launched the joint venture KaMaRetrade in Nizhnekamsk to restore old truck tyres. Marangoni contributed the technology and equipment and organized the training of personnel, while Tatneft provided the production facilities (Neft i kapital 2014b).

Tatneft has received some public recognition for these activities. It was ranked highest among Russian oil companies in the rating of social and environmental responsibility published by the Independent Environmental Rating Agency (NERA): *Ratings of business and regions of Russia: Environmental responsibility and energy efficiency* (Tatneft Press-Tsentr 2010). In September 2017, Tatneft was included in FTSE4Good Emerging Index, which is used to evaluate the activities of companies that demonstrate a commitment to progressive practices in the sphere of environment protection, CSR and corporate governance (Tatneft Press-Tsentr 2017g).

COPING WITH THE CHALLENGES OF CHANGE

Climate Change

According to the *Carbon Disclosure Project Russia 50* report, Tatneft is one of the leading Russian companies concerning climate change issues (CDP 2009), as attested by the company's measures to limit its GHG emissions. For Tatneft, one efficient way of reducing GHG emissions has been the introduction of a system for catching light fractions of hydrocarbons evaporating from storage tanks. Tatneft also seeks to enhance the efficiency of energy use within its Energy-Efficient Economy Programme and has sought to reduce APG flaring (Ilmukova et al. 2013). In fact, it is one of the leaders in the Russian oil sector in terms of APG utilization. In 2015, Tatneft's rate of APG utilization was 95.17% (Tatneft 2018i). In 2018, its level of APG utilization grew to 96.2% (Tatneft 2018i).

Tatneft's APG utilization programme for 2009–13 included the construction of gas power stations, the introduction of furnaces for heating oil that use APG and the construction of gas-gathering systems with the subsequent processing of the gas. In 2012, Tatneft launched a three-year programme to introduce Capstone micro-turbine energy installations that run on APG. In 2013, power stations based on Capstone micro-turbines were commissioned at four Tatneft facilities with a total installed capacity of 4.4 MW (Ilmukova et al. 2013, p. 41). In 2013, Tatneft also introduced APG Control, a system for accumulating and disseminating information on the production and gathering of APG.

Tatneft has also made efforts to promote the use of electric vehicles by commissioning the installation of charging points as part of two projects. In May 2016, Tatneft launched a new facility for charging electric vehicles in Khimgrad industrial park in Kazan. President Minnikhanov took part in the ceremony, studying the specifics of recharging and test driving a smart electric car and stating that electric vehicles were environmentally safe and reliable and that this direction should be developed further (Tatneft press-tsentr 2016d). In June 2017, a second charging facility was opened at Tatneft's fuel station 3 in Almetievsk (Tatneft press-tsentr 2017f).

Sanctions, Crises and Oil Prices

Tatneft weathered the 2008 economic crisis in Russia reasonably well. This does not necessarily mean that the company is good at predicting market changes; nonetheless, it does mean that its strategy represents a robust approach over time. This could be interpreted as indicating that the company is good at handling such changes (Gariffulin et al. 2010).

In 2008, Tatneft successfully pursued a conservative strategy and showed that this could work particularly well for a company holding mostly depleted fields. It was one of the few Russian oil companies that demonstrated production growth during that period; its debt burden remained limited, and it continued generating significant profit. In November 2008, Shafagat Tahautdinov, the then General Director, said that the company intended to cut its investment programme outside of Tatarstan in Orenburg and Samara regions and Nenets Autonomous District while maintaining it at the previous level within Tatarstan (Neft i kapital 2008).

Following the 2014 drop in oil prices, Tatneft's 2015 Annual Report made the following statement:

> Due to the negative macroeconomic factors and fall in oil prices, the company developed a programme of anti-crisis measures, including priority ranking of projects, with due account for the need to maintain full-scale production plans, measures to optimize general and administrative expenses, cut production costs, enhance labour productivity, strengthen control to prevent price hikes by suppliers and improve tender procedures. (Cited in Mukhamadeev et al. 2015)

The current General Director, Nayl Maganov, said in an interview a year after the 2014 oil price collapse that Tatneft's cost cuts did not affect its social programmes. Similar to its policy during the 2008–09 crisis, Tatneft curtailed its activities outside of Tatarstan, making additional investments only to complete commenced projects where it had already invested serious money. Salaries and social benefits were protected as much as possible. In 2015, it even created some 1000 new jobs in the development of extra-viscous oil, constructed new facilities at TANECO and expanded its tyre-manufacturing operations (Zavalishina 2015).

As a 2015 interview with Maganov indicates, he understands that oil prices depend on many factors, including reserves, consumption, geopolitics, demand, technology, innovation and the actions of exchange speculators (Maganov, cited in Zavalishina 2015). Maganov stated that 'one should be oriented towards the price that the market would propose ... Consensus opinion testifies to the growth of prices up to USD 70 by the end of the year. A lot will depend on the events in the Middle East.' The oil price of USD 55 at the time of the interview permitted the company to implement its programmes and to look forward to 15 years of profitable operations. He also reiterated that it would have been better if prices and the exchange rate had remained unchanged since the company had to halt some of its programmes and shelve its boldest ideas.

Maganov admitted in his 2015 interview that the rapid and drastic drop in oil prices affected Tatneft's earnings, and it had to cut costs in spheres that did not directly impact the development of the company (Maganov, cited in Zavalishina 2015). Cost reductions were not applied to investments in important strategic areas, such as oil production, enhanced oil recovery and refining and manufacturing of tyres. Spending was reduced on secondary projects that did not have a critical significance for key production processes:

> Moreover, responding to the challenges of the global market, we developed a series of scenarios depending on the oil price, demand, dollar exchange rate and inflation forecasts. We constantly monitor the situation, and, depending on the current and

strategic analysis, the company uses instruments of the scenario that most fully corresponds to the current market realities. (Maganov, cited in Zavalishina 2015)

Also, in his interview with Kommersant-FM, Maganov said that 'of course, we are not overjoyed that prices were dropping, but we do not see anything lethal for us or unexpected in this process' (Kuzichev and Bogdanov 2016). Moreover, Tatneft's Strategy 2025 contains macroeconomic forecasts of oil prices, the rouble exchange rate and inflation rates and presents three scenarios for the company's market capitalization in 2025. The basic scenario envisages growth of Urals crude oil prices from USD 37 per barrel in 2016 to USD 40 in 2020 and USD 55 in 2025; it also envisages stable fiscal conditions. The negative scenario envisages low oil prices: Urals crude oil blend at USD 24 per barrel in 2016 and USD 45 by 2025. In this case, Tatneft's capitalization would still grow to USD 16 billion. Under the optimistic scenario, the Urals price would be USD 70 per barrel by 2025 and Tatneft's capitalization would increase to USD 28 billion (Tatneft 2016).

According to its 2016 Annual Report, Tatneft was affected by the agreement between Russia and OPEC:

> The global demand for oil is projected to increase in 2017 in the context of the OPEC agreement to curtail oil production. Since January 2017, subject to the agreement between Russia and OPEC, the company committed itself to cut the current production level while preserving all economic and technological aspects to ensure stable operations of oilfields and maintain the production balance. (Voskoboinikov et al. 2016)

International sanctions against Russia in the aftermath of the conflict in Ukraine were not targeted at Tatneft. However, Tatneft was indirectly affected since some of its suppliers encountered serious problems because of the sanctions; for example, one of them delayed deliveries of foreign equipment to TANECO. There were also problems with supplies of pipes for super-viscous oil at high temperatures. However, a Russian company from Perm managed to arrange their manufacture in three months and began to deliver them to Tatneft at reasonable prices.

Maganov noted in 2015 that Tatneft was taking steps aimed at import substitution (Maganov, cited in Zavalishina 2015). Since 2014, it had not been using imported pumps and compressors for its projects; instead, it had been buying pumps and compressors from Russian companies. The company had been placing only domestic orders for reactor and column equipment since 2013. As noted above, the Izhora plant manufactured reactors for the hydrocracking facility that the company launched in 2014 at TANECO (Zavalishina 2015).

CONCLUSION

Tatneft has faced the constant challenge of an ageing and depleting reserve base and has tried to behave proactively by anticipating and responding to changes. For instance, the company began the extraction and production of super-viscous oil to supplement its conventional petroleum reserves. It built a large modern refinery to process its heavy and sour crude; it developed a diverse petrochemical business and expanded into heat and power generation. Tatneft has demonstrated consistent support for domestic companies and capacities and has directed considerable investment to sponsor and maintain the TatNIPIneft research institute.

On climate and CSR, Tatneft does well by the standards of Russian oil companies. In particular, it has demonstrated leadership in the development and promotion of technology to increase energy efficiency and reduce GHG emissions. Tatneft also handled the oil price collapses of 2008 and 2014 well, thanks to its cautious policies. Tatneft has collaborated effectively with international partners and successfully taken home technological and technical expertise to enhance its production and processing operations. However, Tatneft's attempts to develop upstream and downstream activities overseas have been hampered by geopolitics in the Middle East and perhaps by the fact that, unlike its Russian peers, it does not have strong ties with leading politicians in Moscow.

REFERENCES

Astakhova, O. (2017), 'Innovate, cut costs: How a Russian oil firm navigates global supply curbs', accessed 18 September 2017 at https://www.reuters.com/article/us-russia-opec-tatneft/innovate-cut-costs-how-a-russian-oil-firm-navigates-global-supply-curbs-idUSKBN1AP0M7.

Borisov, N. (2006), 'Tatneft ne dozhdalas pomoschi', accessed 12 October 2018 at https://www.vedomosti.ru/newspaper/articles/2006/03/02/tatneft-ne-dozhdalas-pomoschi.

CDP (2009), *Carbon Disclosure Project 2009 Russia 50*, accessed 12 October 2018 at http://www.novatek.ru.

Finam (2019), 'Dobycha Tatnefti v 2018 godu vyrosla na 2%', accessed 28 June 2019 at https://www.finam.ru/analysis/newsitem/dobycha-tatnefti-v-2018-godu-vyrosla-na-2-20190110-102448/.

Finance-Rambler (2018), 'Wintershall ne namerena sotrudnichat s Tatneftyu po razrabotke mestorozhdeniya Dehloran v Irane', accessed 28 June 2019 at https://finance.rambler.ru/business/40209431-wintershall-ne-namerna-sotrudnichat-s-tatneftyu-po-razrabotke-mestorozhdeniya-dehloran-v-irane/.

Gariffulin, I.A., E.A. Tikhturov, V.A. Voskoboinikov et al. (2010), '2010 Annual Report', accessed 14 October 2018 at http://www.tatneft.ru/storage/block_editor/files/5ea3b8fd49bccf5581d7b82da45758a8d045c752.pdf.

Gariffulin, I.G., E.A. Tikhturov, V.A. Voskoboinikov et al. (2012), 'Strengthening positions: Annual Report 2012', accessed 12 October 2018 at http://www.tatneft.ru/storage/block_editor/files/44e0b7c138657f79ff65784e2e97b7c881cb8acd.pdf.

Ilmukova, I.R., R.M. Gareyev, V.Y. Makarov et al. (2013), 'Exploring new horizons: Sustainable Development and Social Responsibility Report 2013', accessed 12 October 2018 at http://www.tatneft.ru/storage/block_editor/files/40831039ac1a8f94071d73055fd76c7b785724c2.pdf.

Interfax (2018), 'Sud Moskvy prodolzhit rassmatrivat isk Tatnefti k Ukraine na $112 mln po delu Ukrtatnafty', accessed 28 June 2019 at https://interfax.com.ua/news/economic/501024.html.

Khisamov, R. (2007), 'Bolshaya istoriya: Dobycha tatarstanskoy nefti nachalas 64 goda nazad', *Neft i kapital*, **8**, 26–31.

Khrennikov, I. and O. Shevelkova (2007), 'Neft na syemi kholmakh', in 'Chyem zakonchitsya korporativnyy konflikt vokrug Moskowskogo NPZ', accessed 12 October 2018 at http://www.compromat.ru/page_21399.htm.kommersant.

Kozlov, D. (2016), 'Igoryu Kolomoyskomu napomnili o dolgye po-angliyski', accessed 12 October 2018 at https://www.kommersant.ru/doc/29069526.

Krivopatre, E. (2017), 'V Almyetyevskye zapustyat sotsialnyy proyekt po aryendye vyelosipyedov', accessed 12 October 2018 at https://www.tatar-inform.ru/news/2017/09/03/570470/.

Kuzichev, A. and O. Bogdanov (2016), 'Nayl Maganov: Neft budet vostrebovana vsegda', accessed 21 January 2019 at https://www.kommersant.ru/doc/2969507.

Lazard Capital Markets (1996), *JSC Tatneft*. Further publication details not available.

Makhnyeva, A. (2016), 'Tatneft, LUKOIL i Novatek voshli v top-10 mirovoy neftegazovoy otrasli', accessed 12 October 2018 at https://www.vedomosti.ru/business/articles/2016/10/25/662257-tatneft-lukoil-novatek.

Malkova, I. and D. Kazmin (2008), 'Isk na $1.1 mlrd', accessed 12 October 2018 at https://www.vedomosti.ru/newspaper/articles/2008/06/07/isk-na-11-mlrd.

Mazneva, E. (2008), 'Spor s novym statusom', accessed 12 October 2018 at https://www.vedomosti.ru/newspaper/articles/2008/06/26/spor-s-novym-statusom.

Mazneva, E. (2010), 'Snova na rynkye', accessed 12 October 2018 at https://www.vedomosti.ru/newspaper/articles/2010/06/29/snova-na-rynke.

Mazneva, E. and I. Malkova (2006), 'S&P nakazalo Tatneft', accessed 12 October 2018 at https://www.vedomosti.ru/newspaper/articles/2006/08/28/sp-nakazalo-tatneft.

Mazneva, E. and A. Tutushkin (2009), 'V Iraq za neftiyu', accessed 12 October 2018 at https://www.vedomosti.ru/newspaper/articles/2009/04/02/v-irak-za-neftyu.

Melnikov, V. (1996), 'Kak reshili – tak i postupyat', accessed 21 January 2019 at https://www.kommersant.ru/doc/242179.

Mukhamadeev, R.N., V.A. Karpov, R.M. Khisamov et al. (2015), '2015 Annual Report of the Tatneft Company', accessed 14 October 2018 at http://www.tatneft.ru/storage/block_editor/files/bc92ca5326966af4b80434bf6d9f7f4455ba4b01.pdf.

Neft i kapital (2004), 'Tatneft', **10**, 136.

Neft i kapital (2008), 'Zhit po sryedstvam', **11**, 38–9.

Neft i kapital (2014a), 'Svetlyy potok: TANEKO zapustila seriyu proizvodstv svetlykh nefteproduktov', **12**, 50–2.

Neft i kapital (2014b), 'Novoye iz starogo', **12**, 54–65.

Oil & Gas Eurasia (2014), 'Russia's Tatneft board discusses shale deposits', accessed 12 October 2018 at https://www.oilandgaseurasia.com/en/news/russia%E2%80%99s-tatneft-board-discusses-shale-deposits.

Panov, A. and G. Gubeydullina (2006), 'America proshchayetsya s Tatneftiyu', accessed 12 October 2018 at https://www.vedomosti.ru/newspaper/articles/2006/11/30/amerika-proschaetsya-s-tatneftyu.

Poussenkova, N. (2010), 'Rossiyskaya neftyanaya promyshlyennost: 20 lyet, kotoryye potryasli mir', accessed 12 October 2018 at http://www.ru-90.ru/node/1319.

Savushkin, S. (1997), 'Aktsii Tatnefti – luchshiy dyebyutant 1996 goda', *Neft i kapital*, **1**, 14–17.

Sberbank CIB Investment Research (2014), 'Tatneft: The conversion of the Tatars', in *Russian Oil and Gas: Two Weddings and a Funeral*, pp. 35–49. Further publication details not available.

Soldatkin, V. (2011), 'Russia Tatneft in $100 mln Libya Capex loss-source', accessed 18 September 2017 at http://www.reuters.com/article/russia-tatneft/russia-tatneft-in-100-mln-libya-capex-loss-source-idUSLDE72R0J120110328.

Takhautdinov, S. (2007), 'Chyetyryokhmilliardnaya tonna nefti Tatarstana mozhyet byt dobyta chyeryez 40–50 lyet', *Neft i kapital*, **8**, 34–8.

Tatar-Inform (2017), 'V Almetyevskye nagradili pobyedityelyey konkyrsa profmasterstva sryedi molodykh rabotnikov Tatnefti', accessed 14 October 2018 at https://www.tatar-inform.ru/news/2017/09/02/570457/.

Tatneft (2016), '20 years of listing on the London Stock Exchange', accessed 12 October 2018 at http://www.tatneft.ru/storage/block_editor/files/911886d0a3712a57c23d26fd79b601010a2562d8.pdf.

Tatneft (2017), 'Osnovnye pokazatyeli', accessed 12 October 2018 at http://www.tatneft.ru/o-kompanii/obshchaya-informatsiya/osnovnie-pokazateli-deyatelnosti-kompanii/?lang=ru.

Tatneft (2018a), 'Obshchaya informatsiya', accessed 11 October 2018 at http://www.tatneft.ru/o-kompanii/obshchaya-informatsiya/?lang=ru.

Tatneft (2018b), 'Struktura aktsionyernogo kapitala', accessed 11 October 2018 at http://www.tatneft.ru/aktsioneram-i-investoram/struktura-aktsionernogo-kapitala/?lang=ru.

Tatneft (2018c), 'Ot pyervogo mestorozhdeniya – do stanovlyeniya kompanii Tatneft (1943–1990)', accessed 11 October 2018 at http://www.tatneft.ru/o-kompanii/istoriya-gruppi-tatneft/ot-pervogo-mestorozhdeniya--do-stanovleniya-kompanii-tatneft-1943--1990/?lang=ru.

Tatneft (2018d), 'Novyeyshaya istoriya (1990–2018)', accessed 11 October 2018 at http://www.tatneft.ru/o-kompanii/istoriya-gruppi-tatneft/noveyshaya-istoriya-19902018?lang=ru.

Tatneft (2018e), 'Razvyedka i dobycha', accessed 12 October 2018 at http://www.tatneft.ru/proizvodstvo/razvedka-i-dobicha/?lang=ru.

Tatneft (2018f), 'Realizatsiya nefti i neftegazoproduktov', accessed 12 October 2018 at http://www.tatneft.ru/proizvodstvo/pererabotka-i-realizatsiya/realizatsiya-nefti-i-neftegazoproduktov/?lang=ru.

Tatneft (2018g), 'Povysheniye effektivnosti neftegazodobychi', accessed 12 October 2018 at http://www.tatneft.ru/proizvodstvo/razvedka-i-dobicha/povishenie-effektivnosti-neftegazodobichi/?lang=ru.

Tatneft (2018h), 'Geographiya neftegazodobychi', accessed 12 October 2018 at http://www.tatneft.ru/proizvodstvo/razvedka-i-dobicha/geografiya-neftegazodobichi/?lang=ru.

Tatneft (2018i), 'Sotsialnyye priorityety', accessed 12 October 2018 at http://www.tatneft.ru/proizvodstvo/pererabotka-i-realizatsiya/neftepererabativayushchee-proizvodstvo/?lang=ru.

Tatneft (2018j), 'Sotsialnyye priorityety', accessed 12 October 2018 at http://www
.tatneft.ru/press-tsentr/press-relizi/more/4639/?lang=ru.

Tatneft Press-Tsentr (2010), 'Tatneft is a "green" company', accessed 12 October 2018
at http://www.tatneft.ru/press-center/press-releases/more/1829/?lang=en.

Tatneft Press-Tsentr (2016a), 'Tatneft vozglavila ryeyting glavnykh innovatorov mira',
accessed 12 October 2018 at http://www.tatneft.ru/press-tsentr/press-relizi/more/
4634/?lang=ru.

Tatneft Press-Tsentr (2016b), 'Tatneft podpisala memorandum o vzaimoponimanii s
Natsionalnoy Iranskoy Neftyanoy Kompaniyey', accessed 12 October 2018 at http://
www.tatneft.ru/press-tsentr/press-relizi/more/4835/?lang=ru.

Tatneft Press-Tsentr (2017a), 'Fitch ratings podtverdilo kryeditnyy reyting Tatnefti na
urovnye "BBB-" so stabilnym prognozom', accessed 12 October 2018 at http://www
.tatneft.ru/press-tsentr/press-relizi/more/5507/?lang=ru.

Tatneft Press-Tsentr (2017b), 'Moody's povysilo kryeditnyy reyting Tatnefti do
Baa3, prognoz-pozitivnyy', accessed 12 October 2018 at http://www.tatneft.ru/press
-tsentr/press-relizi/more/5645/?lang=ru.

Tatneft Press-Tsentr (2017c), 'Tatneft sokhranyayet lidiruyushchiye pozitsii v ryeytinge
the Boston Consulting Group po sozdaniyu stoimosti dlya aktsionyerov', accessed
12 October 2018 at http://www.tatneft.ru/press-tsentr/press-relizi/more/5374/?lang
=ru.

Tatneft Press-Tsentr (2017d), 'TANECO i Chevron Lummus Global LLC dogovorilis
o dalneyshyem sotrudnichestve', accessed 12 October 2018 at http://www.tatneft.ru/
press-tsentr/press-relizi/more/5500/?lang=ru.

Tatneft Press-Tsentr (2017e), 'Apellyatsionnyy sud Anglii i Uelsa yedinoglasnym
ryeshyeniyem tryekh lordov-sudyey postanovil udovlyetvorit v polnom obyemye
apellyatsionnuyu zhalobu PAO Tatneft i otmyenil ryeshyeniye Vysokogo suda
Londona ot 6 noyabrya 2016 g', accessed 12 October 2018 at http://www.tatneft.ru/
press-tsentr/press-relizi/more/5494/?lang=ru.

Tatneft Press-Tsentr (2017f), 'Na AZS Tatnefti otkrylas yeshchye odna stantsiya ely-
ektrozapravki', accessed 12 October 2018 at http://www.tatneft.ru/press-tsentr/press
-relizi/more/5309/?lang=ru.

Tatneft Press-Tsentr (2017g), 'Tatneft voshla v indeks FTSE4Good', accessed 12
October 2018 at http://www.tatneft.ru/press-tsentr/press-relizi/more/5435/?lang=ru.

Tatneft Press-Tsentr (2018), 'Pokazatyeli raboty za dvenadtsat mesyatsev', accessed 11
October 2018 at http://www.tatneft.ru/press-tsentr/press-relizi/more/5608/?lang=ru.

Tatneft Press-Tsentr (2019), 'Sostoyalos godovoye obschee sobraniye aktsionerov
PAO Tatneft', accessed 28 June 2019 at https://www.tatneft.ru/press-tsentr/press
-relizi/more/6483/?lang=ru.

Tatneft Strategy (2016), 'Strategy 2025 Tatneft Group: Key indicators', accessed
12 October 2018 at http://www.tatneft.ru/storage/block_editor/files/d482b7589867
7c4fec75a1d5fdfa7c063fdaff40.pdf.

Tutushkin, A. (2006), 'Tatneft poluchila NPZ na Ukrainye', accessed 12 October 2018
at https://www.vedomosti.ru/newspaper/articles/2006/10/30/tatneft-poluchila-npz
-na-ukraine.

Tutushkin, A. (2008), 'Protsessing poshyel', accessed 12 October 2018 at https://www
.vedomosti.ru/newspaper/articles/2008/01/18/processing-poshel.

Tutushkin, A. and N. Borisov (2006), 'Tatneft stala pyervoy v Rossii', accessed 12
October 2018 at https://www.vedomosti.ru/newspaper/articles/2006/03/27/tatneft
-stala-pervoj-v-rossii.

Tutushkin, A., A. Bausin and A. Nikolskiy (2007), 'Byerkut nalyetyel na Tatneft', accessed 12 October 2018 at https://www.vedomosti.ru/newspaper/articles/2007/10/22/berkut-naletel-na-tatneft.

Voskoboinikov, V.A., R.R. Gaifullina, D.M. Gamirov et al. (2016), '2016 Steady development', accessed 12 October 2018 at http://www.tatneft.ru/storage/block_editor/files/1118d5a842916499d8a97faf3ffb649054b49ebb.PDF.

Zanina, A. and A. Rayskiy (2017), 'Samyye krupnyye sudyebnyye spory 2017 goda', accessed 12 October 2018 at https://www.kommersant.ru/doc/3427410.

Zavalishina, T. (2015), 'Nayl Maganov: Tatneft slishkom dorogo stoit, chtoby yeyo mozhno bylo vzyat i poglotit', accessed 8 September 2017 at https://www.business-gazeta.ru/article/130.

7. Conclusion: champions of change

This book has covered Russia's five largest oil companies: Rosneft, LUKOIL, Gazprom Neft, Surgutneftegas and Tatneft. The chapters dedicated to each of the companies sought to portray the companies and their resilience and ability to adapt to an evolving world.

In the book's introduction, we presented the hypothesis that Russian oil companies are not good at foreseeing and adapting to change. Several subsidiary assumptions supported this hypothesis. First, we conjectured that the unpredictability of post-Soviet society has caused Russian companies to live in the moment and eschew long-term planning, giving rise to a grab-and-run mentality in Russian business culture. Second, Russian companies have subsisted on their Soviet-era assets and infrastructure rather than investing in new exploration and production. Third, the predominance of Soviet-trained, elderly men in the Russian oil company management teams has precluded these companies from responding effectively to the challenges and demands of the contemporary world.

We selected several cross-cutting themes for the exploration of our hypothesis and assumptions in the company chapters. Some of these themes concerned the basic characteristics of the oil companies: organizational transparency, production strategy, corporate social responsibility (CSR), offshore and Arctic oil extraction, innovation and internationalization. Other themes related to the types of change that they had faced and how adept they had been at handling them: oil price fluctuations, sanctions, unconventional oil and climate policy.

In the next section, we provide an overview of the cross-cutting themes we have covered in the company chapters and draw a conclusion on the validity of the hypothesis we presented in the introductory chapter. In the spirit of Karl Popper (1959), our task is to attempt to shoot down our own hypothesis. And, as it turns out, that is exactly what we shall end up doing in this concluding chapter.

CROSS-CUTTING THEMES

Internationalization

Several of the top five Russian oil companies are actively involved in internationalization through cooperation with foreign companies in Russia and investments in other countries. They are also active in overseas upstream and downstream operations. Had there been no sanctions against Russia in connection with the conflict in Ukraine, internationalization would have proceeded even faster. However, the sanctions also encouraged the Russian companies to seek out Asian partners, further diversifying their international ties.

We used the following criteria to profile the internationalization of each company: the number of countries where the company is present (upstream and downstream), foreigners on the board of directors, foreigners on the management board and foreign oil company partners in Russia. Drawing on our review of these criteria for each company, we created Table 7.1 (see p. 184).

Our analysis indicates that of the five oil companies, Rosneft is the most internationalized, followed by LUKOIL, Gazprom Neft, Tatneft and finally Surgutneftegas. However, because of the sanctions, most of Rosneft's Western-based partners have wound down their cooperation with the company, meaning that over time, Rosneft may become less internationalized.

CSR

All the companies have extensive CSR activities, with four of them (Gazprom Neft, LUKOIL, Rosneft and Tatneft) reporting on CSR, while Surgutneftegas publishes environmental reports. Many of the CSR activities take the form of sponsoring local institutions and activities in the regions where the companies operate and are not directly related to the companies' core activities. This form of social support is a remnant of the Soviet system, in which a company was not considered a commercial entity but part of the local government administrative structure with little emphasis on the division of labour and checks and balances but expectations of support for local social infrastructure. Under socialism, oil companies were the founding fathers of oil towns in Western Siberia and had to take care of all aspects of the lives of their workers, including maintenance of waste management facilities, the local steam baths and support for cultural festivals. Many of them divested themselves of social functions when they were privatized, although they maintained some of the Soviet-style traditions because of their symbolic value among employees and local communities.

The different companies covered in this book engage in a wide range of activities that are incorporated under the CSR umbrella and reported on in their sustainability reports. Some of these reports reflect the traditional Soviet approaches, such as support for health facilities and cultural festivals, for example. Others tend to reflect more Western notions of CSR, such as investment in the supply chain capacities of local enterprises. Yet other reports emphasize the rights of local indigenous peoples, civil society organizations and – increasingly – investors. The ways Russian oil companies negotiate benefit-sharing agreements with indigenous peoples combines elements of the Soviet, Russian and international standards and values.

Table 7.1 Internationalization ranking of Russian oil companies

	Gazpr. Neft	Rank	LUKOIL	Rank	Rosneft	Rank	Surgut	Rank	Tatneft	Rank
Number of countries where each company is present (upstream and downstream)	14	3	34	1	25	2	0	5	3	4
Foreigners on the board of directors	None	4.5	4 out of 11	2	7 out of 11	1	None	4.5	2 out of 15	3
Foreigners on the management board	None	2.5	None	2.5	3 out of 11	1	None	2.5	None	2.5
Foreign oil company partners in Russia	Shell, Mubadala, Repsol	2	None	4	BP, ExxonMobil, ENI, Equinor, ONGC, Oil India, Indian Oil, Bharat PetroResources, Sinopec, CNPC, ChemChina Beijing Gas, Pertamina	1	None	4	None	4
Average rank		3		2.38		1.25		4		3.38

Methodological note: Companies are ranked from 1 (most international) to 5 (least international). Where they are identical, we have given them the same rank.

Oil Price Fluctuations

Due to oil price instability, the oil industry is one of the most changeable in the world, with extreme highs and lows. Oil price fluctuations are, therefore, the most important type of change with which any oil company has to grapple. Russian oil companies have handled the price hikes and collapses of the past 20 years surprisingly well. For the purpose of our analysis, the oil price collapse of 2014 is particularly interesting, as it coincided with Western sanctions targeting the Russian petroleum sector over Russia's involvement in the conflict in Ukraine. It was the perfect storm, and the result says something about Russian oil companies: They weathered the storm.

There are several reasons for the success of Russian oil companies in handling oil price fluctuations. The points listed here mostly concern the context in which the companies operate rather than relating to the Russian oil companies as organizations in their own right. First, even when oil prices were above USD 100 per barrel, for the Russian oil companies, these prices did not, in practice, exceed USD 35–40 because of the Russian revenue-based tax system (see the chapter on Gazprom Neft).

Second, as oil price fluctuations follow a cyclical pattern, conservative behaviour on the part of the oil companies can help them smooth out developments (not being too speculative when prices are high and not employing harsh cutbacks when they are low). The downside is that during a downturn, the Russian oil companies do not enjoy the same efficiency gains as their Western counterparts. The upshot is that the Russian oil companies are stable and resilient to price fluctuations but not world leaders.

Third, while the close linkage between the exchange rate of the Russian rouble and the oil price are detrimental to other parts of the Russian economy, it is highly beneficial for the Russian oil companies. When the oil price collapses, the rouble also tends to lose much of its value, greatly reducing their expenses.

Finally, after both the 2008 and 2014 oil price collapses, the state stepped in to provide financial aid to several companies. This was accomplished by reducing taxes and easing licensing and general pressure on the companies, on the one hand, and facilitating their access to capital, on the other.

Sanctions

Russia's oil majors also succeeded in coping with the 2014 Western sanctions over the conflict in Ukraine. Except for Tatneft, all the other companies covered in the book were targeted by the sanctions, and above all Rosneft. The companies' response to sanctions was sensible and included import substitution, a turn to new foreign partners in Asia and requests for financial assistance from the state. While the Russian oil companies suffered but survived the sanctions, the Chinese and Indian companies actually benefited from them since they managed to gain access to Russian assets at attractive prices.

Joint projects in the Arctic and offshore and unconventional oil that had been planned for implementation with Western-based companies were shelved. However, the projects that did not involve Western partners, such as Gazprom Neft's Arctic Prirazlomnoye field in the Pechora Sea and the work of Surgutneftegas and Tatneft on the Bazhenov shale formation in Western Siberia and on extra-viscous oil in Tatarstan, were carried out according to plan.

Overall oil production was little impacted by the sanctions. Figure 7.1 indicates a slight slowdown of oil production after the onset of the first wave of sanctions in early 2014. However, this may just as well have been due to the concurrent oil price collapse, and later also due to coordination between OPEC and Russia to cap output. Whatever caused Russian oil production growth to stagnate, it is clear that sanctions did not cause it to collapse. And by the end of 2018 it started rising again. A significant part of the reason for this can be found in Russia's oil majors' resilience to change. As oil price fluctuations are the most important type of change for oil companies, this undermines the hypothesis that Russian oil companies are bad at coping with change.

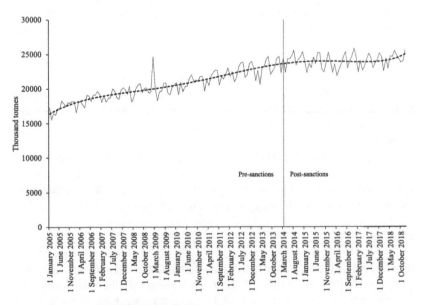

Figure 7.1   Russian oil production and sanctions

Source: GKS (2019); MinEnergo (2019).

Onshore and Offshore

Until the imposition of sanctions in 2014, the Russian petroleum sector was a story of a landlubber trying to go to sea. The Russian oil and gas industry had a long history of onshore operations but with limited offshore experience. As the major oilfields discovered in the 1960s and 1970s were becoming depleted, the Russian oil companies tried to access Russia's offshore resources, which were expected to be significant, especially in the Arctic. To this end, they needed help from the international oil companies. Thus, the directing of sanctions at Russian offshore and Arctic developments can be interpreted as an attempt to hit where it hurts. In this regard, the sanctions were largely successful, especially in the case of the highly promising Kara Sea, which had to be shelved for the time being. However, Russia is a vast country and also has many onshore resources that can be exploited.

This short-term outlook aside, it takes a long time to discover and develop an oilfield, and in practice, the impact of the sanctions may not be felt for another decade or so after they were introduced in 2014 (Fjaertoft and Overland 2015). This situation may appear paradoxical mainly because the intention behind the sanctions was to punish Russia swiftly and severely for its role in Ukraine. However, as the Russian oil industry continues to evolve, locates new partners in China, India and other countries and sees an erosion of the will to uphold sanctions, it is unclear whether the sanctions will have a significant impact in the long run either. It is also unclear whether the new Asian partners can meet Russia's needs as successfully as the Western oil companies, given the profound international experience and cutting-edge petroleum technology of Western majors.

Innovation

Innovation is an important aspect of the ability to respond to the challenges of a changing world. President Putin has repeatedly called on the major Russian companies for greater innovation (Putin 2017), and Dmitry Medvedev made innovation the centrepiece of his presidency (Overland 2011). Nonetheless, there is a lingering impression that Russian oil companies are not innovative.

Although the Russian oil companies are far from being world leaders in petroleum technology, they have been trying to advance technologically. Among the five companies covered in this book, Gazprom Neft exhibits the greatest interest in innovation, including technologies to enhance well productivity, the development of the Bazhenov shale formation, oil recovery from depleted fields, manufacture of catalysts for refining and digitalization of oil production (see the chapter on Gazprom Neft).

Concerning unconventional oil, two other Russian oil companies have innovative profiles: Surgutneftegas has a long record of extracting unconventional oil from the Bazhenov formation, and Tatneft seeks to gain as much as possible from its super-viscous fields (see the chapters on Surgutneftegas and Tatneft). However, a perennial challenge for the development of unconventional oil in Russia has been the country's extensive resources but limited effort to extract them. As the conventional resources in Western Siberia decline and new discoveries become less frequent, more companies may eventually find that they are forced to learn how to extract unconventional oil.

Climate Change and Energy Policy

In the long term, the most challenging change faced by oil companies – Russian or international – may be that of climate policy. Most of the five companies covered in this book have, at best, an ambivalent attitude towards climate change. In the terminology of Sherlock Holmes, this is a case of dogs that did not bark. It was difficult for us to find data on the positions of the companies on climate change, and this difficulty is in itself a sign that they are not proactive in this area. Their main real contribution, if any, is limiting the flaring of associated petroleum gas, which is required by the Russian authorities in order to maximize the use of Russia's natural resources.

However, a slight tendency for greater attention to climate change among the companies can be discerned in both their public statements and various publicly available company reports (see the chapter on Rosneft). Nonetheless, similar to many of their foreign peers, the Russian oil companies still believe that fossil fuels are here to stay.

Of all the companies, LUKOIL leads on reporting on greenhouse gas (GHG) emissions, utilization of associated petroleum gas, introduction of energy efficiency measures and preliminary steps concerning renewable energy, mostly outside Russia. Gazprom Neft has been similarly positioned, although in 2018 its utilization of associated petroleum gas declined again. In this regard, they follow the same trajectory as their international peers, lagging perhaps 10–15 years behind European companies, such as Equinor, Shell and Total but only a few years behind the American corporations.

Transparency

We found that most of the Russian oil companies are relatively transparent in their operations. This is especially true of those that are listed on foreign stock exchanges and thus required to share detailed financial information. We assessed the transparency of the five companies in several areas: recent regular international reserve audit, recent regular reporting under international accounting standards (IAS), clear public information on shareholders, information on share ownership by members of the board of directors and/or top managers, operating data on fields and wells, sustainability reports in accordance with the Global Reporting Initiative (GRI), earliest financial annual report and number of press releases in 2018.

Drawing on our review of these areas, we ranked the transparency of the five companies and found that most of them were very transparent (Table 7.2).

Table 7.2   *Transparency ranking of Russian oil companies*

	Gazp. Neft	LUKOIL	Rosneft	Surgut	Tatneft
A. Recent regular international reserve audit	1	1	1	0	1
B. Recent regular IAS statements	1	1	1	1	1
C. Clear data on shareholders	1	1	1	0	0.5
D. Data on share ownership by the board of directors' members and/or top managers	1	1	1	1	1
E. Operating data on fields and wells	1	1	0	0	0
F. Sustainability reports in accordance with GRI	1	1	1	0	0
G. Earliest financial annual report and number of press releases in 2018	0.95	0.7	0.59	0.07	0.82
H. Overall score	5.70	4.20	2.95	0.14	2.87

Notes:
A–F: 1 = yes, 0 = no
G: Earliest annual financial report available on the website. Number of press releases in 2018: normalized and averaged.
H: (sum of A–F) * G

MASTERS OF ADAPTATION

We hypothesized that Russian oil companies might not be good at foreseeing and handling change but in fact they have dealt with more of it – and done so no less successfully – than their Western counterparts. In the two decades from 1998 to 2018, Russian oil companies had to contend with the oil price drop and the collapse of the Russian economy in 1998, the gangster capitalism of the 1990s and the authoritarian capitalism of the 2000s and 2010s. The arrest of the country's once richest oilman, Mikhail Khodorkovskiy, in 2003 had a considerable impact; as did Rosneft's takeover of four of the best known Russian oil companies. The global financial meltdown of 2008 and the ensuing collapse of the oil price caused further chaos. This was followed by mass demonstrations against the Russian government in 2011–12; the conflict in Ukraine in 2014 and the resulting Western sanctions against the Russian petroleum industry; and the collapse of the oil price and the Russian rouble in 2014. Meanwhile, the Soviet Union's former enemy, China, rose to become the world's biggest gross oil importer in 2018, in the context of the growing international isolation of Russia and the deterioration of its relations with the Western world.

Obviously, several of these developments also had implications for the Western oil companies; however, the Russians had to deal with a whole raft of additional domestic and international issues which did not concern their Western counterparts. In addition, the Western oil companies did not prove any better at predicting the future than their Russian counterparts; none of them foresaw the developments in Ukraine and their consequences for the relationships between Russia and the West. The Russian oil companies that survived the 1990s learned how to change with changing times and adapt to the new rules of the game when Vladimir Putin came to power and launched the étatization of the economy (Godzimirski 2018).

Having examined Russia's five largest oil companies and how they respond to a changing environment, we find that they are in fact masters of change. The simultaneous oil price collapse and introduction of sanctions in 2014 might have been enough to wipe them out, but the companies appear to have escaped more or less unscathed. Some of the points presented as part of the hypothesis in the introduction to this book – that Russian companies are characterized by short-termism and a grab-and-run mentality, that they thrive on assets created during the Soviet period and are unwilling or unable to invest in new developments, that Russian society is fundamentally conservative – are challenged by our findings.

Also as a people, one cannot say that Russians are strangers to change. Over the past century and a half, they have faced more dramatic changes than most peoples, including the abrupt transition from the conservative tsarist regime to Soviet communism and Stalinism, two world wars which were particularly bloody for Russia, and the collapse of the Soviet Union and its entire socio-economic system.

SURVIVAL OF THE FITTEST

A possible explanation for the resilience of Russia's oil majors is that they are the survivors of the cut-throat post-Soviet period, which only the fittest survived. Many companies were gobbled up: Bashneft, Itera, Sibneft, TNK-BP and YUKOS, to name a few. Also, many of the most powerful oilmen of the early post-Soviet period, such as Victor Ageev, Victor Gorodilov, Sergei Muravlenko and Anatoliy Fomin, had to concede defeat. By contrast, Vagit Alekperov (see the chapter on LUKOIL) and Vladimir Bogdanov (see the chapter on Surgutneftegas) were strong and able to adapt to change, while Alekperov even anticipated and initiated some change himself. The Russian oil industry is mostly about people – larger-than-life personalities who know how to move with the times, not only adapting to, but in some cases writing the new rules of the game.

In her 2008 study of organizational transformation in the Russian oil industry, Dixon (2008) somewhat paradoxically admired TNK-BP and YUKOS for their innovativeness, openness to external impulses and efficiency while casting LUKOIL and Surgutneftegas as inefficient and authoritarian relics of Soviet management culture. Today, LUKOIL and Surgutneftegas are doing fine while TNK-BP and YUKOS have been cannibalized by their peers. Clearly, LUKOIL and Surgutneftegas had something that TNK-BP and YUKOS lacked: the ability to cope with a changing environment not through innovation and revolution but through political savviness, discipline, stability and stoicism. What Western commentators have interpreted as stagnation and outdatedness may have turned out to be strengths in Russia's post-Soviet business environment. LUKOIL and Surgutneftegas understood what was needed to survive and prosper in Russia, and did so better than YUKOS.

Dixon (2008, p. 41) writes: 'Traditional Russian organisations are characterised by a command-and-control approach which is manifested in authoritarianism, obedience to authority, the use of coercive power, and an emphasis on rank and status.' This point finds strong support in the broader literature (see Henry and Sundstrom 2007; Husted et al. 2012; Kets de Vries 2001; Kogut and Zander 2000; Kornai 1992; Kryukov and Moe 2018; McCarthy et al. 2005; Vlachoutsicos and Lawrence 1996).

Also, we find that Russian oil and gas companies are characterized by a highly hierarchical, top-down structure. Employees fear making errors, being blamed for errors, or being swept out in crackdowns or changes of management. Disobedience and failure to carry out orders are often punished harshly. Staff may not only be fired by their current employer, but de facto banned from working for any company in the industry. Such a culture of fear is not conducive to organizational foresight. Thus, one possible interpretation is that, due to hierarchies and discipline, Russian oil companies are indeed not very good at foreseeing changes, but the same discipline makes them robust and able to handle changes once they occur. The fact that they have a lot of practice dealing with dramatic changes may also make them skilled at it.

A NOTE OF CAUTION: RUSSIA HAS NO TESLA

In this book, we have focused on Russia's five main oil companies because of the decisive impact of their performance on Russia's petroleum sector and hence the performance of the Russian economy and the strength of the Russian state. Although we have ended up granting Russian oil majors a clean bill of health as oil companies, this does not mean that Russia and the Russian petroleum sector are immune to the changes in international oil demand arising from climate policy. Russia may have well-managed oil companies; however, it hardly has many things other than these companies; in other words, it has few other companies positioned to take on an important role in a decarbonized world.

Western actors are not necessarily better at predicting the future, but they are more diverse. For example, the two American companies ExxonMobil and Tesla embody diametrically opposite visions of where the world is headed. This means that whichever vision is right, the United States, as a country, has a bet on it. Russia's bets, however, are less diversified, and that is a vulnerability.

REFERENCES

Dixon, S. (2008), *Organisational Transformation in the Russian Oil Industry*, Cheltenham, UK and Northampton, MA, USA: Edward Elgar.

Fjaertoft, D. and I. Overland (2015), 'Financial sanctions impact Russian oil, equipment export ban's effects limited', *Oil & Gas Journal*, **113** (8), 66–72, accessed 25 April 2020 at https://www.researchgate.net/publication/281776234_Financial_Sanctions_Impact_Russian_Oil_Equipment_Export_Ban's_Effects_Limited.

GKS (2019), 'Prosmyshlennoe proizvodstvo', accessed 27 January 2019 at http://www.gks.ru/wps/wcm/connect/rosstat_main/rosstat/ru/statistics/enterprise/industrial/#.

Godzimirski, J. (2018), *The Political Economy of Russian Aluminium: Between the Dual State and Global Markets*, London: Palgrave Macmillan.

Henry, L.A. and L.M. Sundstrom (2007), 'Russia and the Kyoto Protocol: Seeking an alignment of interests and image', *Global Environmental Politics*, **7** (4), 47–69.

Husted, K., S. Michailova, D.B. Minbaeva and T. Pedersen (2012), 'Knowledge-sharing hostility and governance mechanisms: An empirical test', *Journal of Knowledge Management*, **16** (5), 754–73.

Kets de Vries, M.F.R. (2001), 'The anarchist within: Clinical reflections on Russian character and leadership style', *Human Relations*, **52** (5), 585–627.

Kogut, B. and U. Zander (2000), 'Did socialism fail to innovate? A natural experiment of the two Zeiss companies', *American Sociological Review*, **65** (2), 169–90.

Kornai, J. (1992), *The Socialist System: The Political Economy of Communism*, Oxford and New York: Oxford University Press.

Kryukov, V. and A. Moe (2018), 'Does Russian unconventional oil have a future?', *Energy Policy*, **119**, 41–50.

McCarthy, D., S. Puffer, O. Vikhanski and A. Naumov (2005), 'Russian managers in the new Europe: Need for a new management style', *Organizational Dynamics*, **34** (3), 231–46.

MinEnergo (2019), 'Statistika', accessed 27 January 2019 at https://minenergo.gov.ru/activity/statistic.

Overland, I. (2011), 'Modernization after Medvedev?', *Russian Analytical Digest*, **105**, 2–4.

Popper, K. (1959), *The Logic of Scientific Discovery*, London: Hutchinson.

Putin, V. (2017), 'O chem rasskazal Vladimir Putin ya plenarnom zasedanii PMEF', *Rossiyskaya Gazeta*, accessed 20 November 2017 at https://rg.ru/2017/06/02/reg -szfo/o-chem-rasskazal-vladimir-putin-na-plenarnom-zasedanii-pmef.html.
Vlachoutsicos, C.A. and P.R. Lawrence (1996), 'How managerial learning can assist economic transformation in Russia', *Organization Studies*, **17** (2), 311–25.

Appendix 1: Corporate history of Rosneft

1991	The Ministry of the Oil and Gas Industry of the USSR was abolished, and Rosneftegaz was formed based on it (Ppnf.ru 2011).
1993	Alexander Putilov, the former Head of Uraineftegaz, was appointed President of Rosneft (Lenta.ru 2012; Osipov 2013).
1993	The state enterprise Rosneft was founded, based on assets previously held by Rosneftegaz and Rosnefteprodukt (Osipov 2013; Rosneft 2017l).
1993–94	Rosneft lost valuable assets to the newly established SIDANCO, Eastern Oil Company and ONACO.
1994	Purneftegaz, the key subsidiary of Rosneft, was transferred to SIDANCO.
1995	Rosneft lost valuable assets to the newly established TNK and Sibneft.
1995	Open joint stock company (OJSC) Rosneft was established in accordance with Russian Government Decree 971 (Rosneft 2017l).
1995	Purneftegaz was transferred back to Rosneft.
1996	The production-sharing agreement (PSA) for the Sakhalin-1 project came into effect. Its shareholders included Exxon as project operator (30%) and SODECO (30%), with the remaining 40% being divided between Sakhalinmorneftegaz-Shelf (a subsidiary of SMNG) and Rosneft.
1996	SIDANCO filed a lawsuit demanding the return of Purneftegaz.
1997	The court recognized the validity of SIDANCO's claim concerning Purneftegaz, but Vladimir Potanin decided to return it to Rosneft.
1997	Yuri Bespalov, the former Minister of Industry (1996–97), was appointed President of Rosneft (Lenta.ru 2012).
1997	The government developed the first plans to privatize Rosneft.
1998	The government decided to sell 75% +1 share of Rosneft. The auction was first scheduled for May, but it was then postponed until autumn.
1998	The financial crisis hit; the privatization auction was postponed indefinitely.
1998	Sergei Bogdanchikov, the former General Director of Rosneft's subsidiary Sakhalinmorneftegaz, was appointed President of Rosneft (Lenta.ru 2012; Osipov 2013).
1999	A cost-cutting programme was launched (Sputniknews 2005).
2000	For the first time after the financial crisis, Rosneft reported a profit (Lenta.ru 2012).
2000	Rosneft started rebuilding Chechnya's oil industry (Lenta.ru 2012).
2000	Rosneft received permission from the Russian Anti-Monopoly Ministry to increase its stake in subsidiaries to 75%.

2001	An agreement was signed with Gazprom on joint Arctic offshore development, and the Sevmorneftegaz joint venture was established on a 50:50 basis (Sputniknews 2005).
2001	Rosneft became the first Russian oil company after the 1998 crisis to issue Eurobonds to European and American investors (Sputniknews 2005).
2001	Rosneft and Stroytransgaz were declared the winners of the first international tender for exploration of hydrocarbons in Algeria.
2001	Rosneft sold half of its share in Sakhalin-1 to the Indian Oil and Natural Gas Corporation (ONGC).
2001	Rosneft and Sinopec began working on a 50:50 basis on the Adaisk zone in Kazakhstan.
2002	Rosneft obtained a licence to develop the Kaygansko-Vasyukansky block as part of the Sakhalin-5 project (Rosneft 2017l).
2003	Rosneft bought Severnaya Neft (Lenta.ru 2012).
2003	Rosneft took under its control the Vankor field in Eastern Siberia.
2003	Rosneft bought Selkupneftegaz and Pur Oil Company (Novatek 2005).
2003	Rosneft bought a 50% stake in the Polar Lights joint venture with ConocoPhillips.
2003	Rosneft received a licence to develop the Veninsky block, part of the Sakhalin-3 project (Rosneft 2017l).
2003	Rosneft received an exploration licence for the West Kamchatka shelf.
2003	In February, at a meeting between President Vladimir Putin and representatives of the Russian Union of Industrialists and Entrepreneurs, the Head of YUKOS, Mikhail Khodorkovskiy, confronted the Head of Rosneft, Sergei Bogdanchikov (Lenta.ru 2012).
2003	First, Platon Lebedev and then Mikhail Khodorkovskiy were arrested, and the YUKOS case began.
2004	Vladimir Putin's close ally Igor Sechin was appointed chairman of the board of directors of Rosneft (Osipov 2013; Rosneft 2017l).
2004	Rosneft signed a MoU with KNOC concerning the joint development of the West Kamchatka shelf (Rigzone 2004).
2004	In September, Vladimir Putin approved a government proposal to fully incorporate Rosneft into Gazprom in exchange for 10.74% in the gas monopoly that was held by its subsidiaries.
2004	The government froze the shares of YUKOS subsidiary Yuganskneftegaz, claiming that shares would soon be sold to cover tax debts (Lenta.ru 2012).
2004	In December, the Yuganskneftegaz auction was held; the company was bought by Baikal Finance Group.
2004	Rosneft acquired Baikal Finance Group (NGRF 2005; Osipov 2013).
2005	Chinese banks lent Rosneft USD 6 billion to be repaid in the form of oil exports to China.
2005	Oil production started in the Sakhalin-1 project (Sputniknews 2005).
2005	Rosneft signed an agreement with the Chinese company Sinopec to conduct geological exploration and prospecting within Veninsky block of Sakhalin-3 (China Daily 2007).
2005	The Gazprom-Rosneft merger was disbanded (Wikipedia 2018).

2005	Rosneft bought 25.9% of Verkhnechonskneftegaz from Interros, becoming a partner with TNK-BP (Reuters 2007).
2005	Rosneft sold its stake in Sevmorneftegaz to Gazprom.
2005	A PSA and an agreement on joint activities were signed in Astana for the Kurmangazy block to be developed by Rosneft and Kazmunaigaz in Kazakhstan.
2006	Rosneft nominated independent members to its board of directors.
2006	Rosneft made an initial public offering (IPO) on the London Stock Exchange and the Russian Trading System (RTS), selling 14.8% of its shares (Rosneft 2017).
2006	Rosneft acquired the debts of YUKOS, becoming its second-biggest debtor after the Russian tax authorities.
2006	Sinopec bought Udmurtneft, acting in the interests of Rosneft.
2006	Rosneft and CNPC signed a cooperation agreement during Vladimir Putin's visit to China and established Vostok Energy.
2007	Rosneft decided to raise a loan of up to USD 13 billion and provide a guarantee for Rosneft-Development to raise up to USD 9 billion funds in debt to acquire the oil assets of YUKOS (NGFR 2005).
2007	Rosneft bought YUKOS's oil assets during the bankruptcy auctions of the latter (Osipov 2013).
2007	Rosneft was included in the list of strategic enterprises.
2007	Rosneft (49%) and CNPC (51%) established the PetroChina-Rosneft Orient Petrochemical (Tianjin) Company to build a refinery in China.
2008	Igor Sechin was appointed Deputy Prime Minister in charge of the energy sector.
2008	President Dmitriy Medvedev signed the federal law on 'Introducing Amendments to the Federal Law on the Continental Shelf', under which only Gazprom and Rosneft were given access to offshore fields.
2008	Rosnedra refused to extend the West Kamchatka offshore exploration licence of Rosneft and its Korean partners.
2008	Rosneft, LUKOIL, Gazprom Neft, TNK-BP and Surgutneftegas established the National Oil Consortium to develop Junin-6 block in Venezuela.
2009	Rosneft launched commercial production at Vankor (Rosneft 2017l).
2009	Chinese banks lent Rosneft USD 15 billion to be repaid by oil exports to China (Reuters 2009).
2010	Rosneft acquired stakes in four refineries in Germany from Petróleos de Venezuela (PDVSA) (Rosneft 2017l).
2010	Sergei Bogdanchikov was fired from Rosneft while he was on a business trip. Eduard Khudainatov, Head of Severneftegazprom, replaced Bogdanchikov (Lenta.ru 2012; Osipov 2013).
2010	Opposition politician and activist Aleksei Navalny launched a campaign for greater transparency. Navalny tried to find out the details of the contract between Rosneft and CNPC (Osipov 2013).
2010	Rosneft planned to build an oil refinery in Grozny, Chechnya (Marchmonthnews 2010).
2010	Rosneft (49%) and Crescent Petroleum (51%) from the United Arab Emirates launched a project to drill for gas in Sharjah.

2010	Rosneft invited Chevron to participate in its Black Sea ventures.
2011	Chevron withdrew from the joint Black Sea project with Rosneft.
2011	President Medvedev announced that all vice premiers and ministers of the Russian government must withdraw from the boards of directors of state-owned companies. Igor Sechin resigned from the board of directors of Rosneft and was replaced by the Vice President of the Russian Academy of Sciences, Alexander Nekipelov (Lenta.ru 2012).
2011	A strategic alliance between Rosneft and BP was announced but it was soon torpedoed by the Russian shareholders of TNK-BP, that is, the AAR consortium (Rosneft 2011).
2011	Rosneft formed a strategic partnership with ExxonMobil, centred on the joint development of Russian offshore Arctic oil and oil in the Mexican Gulf and Canada (Lenta.ru 2012).
2011	The Kazakh Ministry of Oil and Gas terminated the Adaisk block PSA at the request of Rosneft and Sinopec.
2012	In March 2012, Rosneft acquired 35.3% of Taas-Yuryakh Neftegazodobycha from Sberbank and also increased its stake from 70.78% to 99.87% in Vostsibneftegaz (Rosneft 2012a).
2012	Rosneft signed an agreement on strategic cooperation with Itera (Rosneft 2012b).
2012	Igor Sechin became CEO of Rosneft (Osipov 2013; Rosneft 2017l).
2012	Rosneft reached a binding agreement to take over TNK-BP. The deal was finalized in March 2013 (Katona 2016; Osipov 2013; Rosneft 2017l).
2012	Rosneft and ExxonMobil signed an agreement to cooperate on tight oil reserves at the Bazhenov and Achimov formations in Western Siberia (Rosneft 2017l).
2012	Rosneft signed an agreement with Statoil (now Equinor) to establish a joint venture to work in the Barents Sea and the Sea of Okhotsk (Rosneft 2017l).
2012	Rosneft and ENI formed a joint venture to develop licence areas in the Barents and Black Seas (Rosneft 2017l).
2012	Rosneft and its international partners signed joint declarations on sustainable Arctic offshore developments.
2012	Rosneft signed a major deal on gas supplies with Inter RAO.
2012	Rosneft and Itera Group established a joint venture based on the assets of NGK Itera (Rosneft 2012c).
2012	RN Cardium Oil bought a 30% share in the Cardium project in the province of Alberta from ExxonMobil.
2012	Rosneft, Transneft and CNPC resolved their conflict and agreed upon new delivery terms with the Russian companies providing a discount of USD 1.5 per barrel to CNPC.
2013	Rosneft closed the deal on the acquisition of TNK-BP (Rosneft 2017l).
2013	Rosneft reached an agreement to buy several petroleum assets of ALROSA (Rosneft 2013g).
2013	Rosneft reached an agreement with Enel to buy 40% in Arctic Russia B.V. (Rosneft 2013a).
2013	Rosneft and Novatek reached an agreement on an asset swap (Rosneft 2013b).
2013	The Federal Anti-Monopoly Service allowed RN-East Siberia to purchase the remaining 64.67% of Taas-Yuryakh Neftegazodobycha (Rosneft 2013c).

2013	Rosneft and ExxonMobil expanded their strategic cooperation by including seven additional licensing plots covering an area of 600 000 square kilometres in the Chukchi, Laptev and Kara seas (Rosneft 2013d).
2013	Rosneft and CNPC signed an export contract to deliver 360 million metric tonnes of crude over 25 years, worth some USD 270 billion. In addition, Rosneft and Sinopec signed a contract worth USD 85 billion under which Rosneft would supply 100 million metric tonnes of crude over ten years (Reuters 2013).
2013	The dividend was increased to RUB 85 billion, and RUB 2.7 trillion was transferred to the budget of the Russian Federation (Rosneft 2017l).
2013	Rosneft, Gazprombank, Sovkomflot and DSME signed a MoU on cooperation for creating a shipbuilding and industrial cluster in Primorsky Kray.
2013	Rosneft was officially charged with the construction of an oil refinery in Chechnya.
2013	Rosneft became the third-largest gas producer in Russia after Gazprom and Novatek.
2013	Rosneft and ExxonMobil made plans to build an LNG plant in the Far East of Russia.
2013	Rosneft and Novatek successfully lobbied for changes in gas exports: a law granting export rights to certain LNG projects came into force, essentially ending Gazprom's monopoly.
2013	Rosneft (40%) and CVP, a subsidiary of PDVSA, signed an agreement establishing Petrovictoria S.A. to develop Carabobo 2.
2013	Neftegaz America Shelf LP, a subsidiary of Rosneft, bought a 30% share in 20 of ExxonMobil's deep-water blocks in the American part of the Mexican Gulf (Rosneft 2013e).
2013	Rosneft and Statoil (now Equinor) successfully participated in the twenty-second licensing round held by the government of Norway.
2013	Rosneft obtained important gas assets in Vietnam through its acquisition of TNK-BP.
2013	Rosneft bought a stake in the Saras refinery in Italy (Rosneft 2013f).
2014	Rosneft together with ExxonMobil discovered a large oil and gas field in the Kara Sea and called it Pobeda (Victory) (Rosneft 2017l).
2014	Rosneft strengthened its foothold in Germany by acquiring Total's 16.67% stake in the Schwedt refinery (Katona 2016).
2014	Rosneft increased its stake in the Venezuelan National Oil Consortium to 80% (Rosneft 2014e).
2014	Rosneft and PDVSA signed an agreement on cooperation to start production in the Rio Caribe and Mejillones blocks in Venezuela (Processing 2014).
2014	Rosneft and PDVSA signed their first long-term contract for deliveries of oil and petroleum products to Russia from Venezuela.
2014	Rosneft bought 100% of the Orenburg Drilling Company from VTB Leasing (Rosneft 2014a).
2014	US sanctions were introduced against Rosneft and against Igor Sechin personally.
2014	17 Rosneft subsidiaries were placed under US sanctions, and ExxonMobil withdrew from the joint Kara Sea exploration drilling project (Katona 2016).
2014	Rosneft bought drilling and well-repair assets in Russia and Venezuela from Weatherford International.
2014	Rosneft and Cuban CUPET signed a Memorandum of Cooperation to implement projects in Cuba (Rosneft 2014b).

2014	Rosneft began to increase the share in the Solimoes project in Brazil, which it had inherited from TNK-BP (Rosneft 2014c).
2014	Rosneft acquired the Bishkek Oil Company in Kyrgyzstan (Rosneft 2014d).
2015	Rosneft commenced oil production at the Arkutun-Dagi field using Berkut, the world's largest drilling platform (Rosneft 2017l).
2015	Rosneft acquired the SANORS holding company (Novokuybyshevsk Petrochemical Company) (Rosneft 2017l).
2015	Rosneft bought the Russian subsidiary of Trican Well Service (Rosneft 2015d).
2015	The joint venture between Rosneft and BP to exploit the Domanic formation in the Orenburg region halted its activities because of the sanctions (Industryruss 2015).
2015	Moody's downgraded Rosneft's rating from Baa2 to Baa3, and S&P downgraded its long-term credit rating of Rosneft's foreign currency debt from BBB- to BB+, with a 'negative' outlook.
2015	Rosneft closed a deal for the sale of its 50% in the joint venture Polar Lights with ConocoPhllips (Rosneft 2015a).
2015	Rosneft and BP signed a deal to cooperate within the framework of Taas-Yuryakh Neftegazodobycha (BP bought 20%) (Rosneft 2015b).
2015	Rosneft and ExxonMobil were announced the winners of three blocks in Mozambique, with ExxonMobil as the project operator.
2015	Rosneft and Sinopec signed an agreement on the joint development of the Russkoye and Yurubcheno-Tokhomskoye fields (Rosneft 2015c).
2016	Rosneft finalized a deal to sell 19.5% of its stock to Glencore and the Qatari sovereign wealth fund (Golubkova et al. 2016).
2016	Rosneft overtook Gazprom in terms of its capitalization in RTS.
2016	Rosneft bought the Bashneft oil company from the state for RUB 330 billion.
2016	Rosneft bought the Targin service company from AFK Sistema (Rosneft 2016d).
2016	Rosneft (51%) and BP (49%) signed an agreement on the establishment of the Yermak Neftegaz joint venture (Rosneft 2016a).
2016	Rosneft and ExxonMobil returned their licences for the blocks in the Gulf of Mexico.
2016	Rosneft dissolved its Ruhr Oel partnership with BP in Germany (Rosneft 2016b).
2016	Rosneft sold 49.9% of Vankorneft to India's ONGC and a consortium of Indian investors, and 29.9% of Taas-Yuryakh Neftegazodobycha to the consortium (TASS 2016).
2016	In November, Alexei Ulyukayev, the Minister of Economic Development, was detained in the offices of Rosneft on charges of extortion and taking a bribe of USD 2 million for allowing the deal to sell shares of Bashneft to go through (BBC 2017; Caroll 2017; Revonenko 2017).
2016	Rosneft and Statoil (now Equinor) drilled two dry wells in the Sea of Okhotsk (OffshoreEnergy 2016b).
2016	Rosneft and the Indonesian company Pertamina signed preliminary agreements on the sale to the Indonesian company of stakes in the Russkoye field (up to 37.5%) and in the northern part of Chaivo (up to 20%) (OffshoreEnergy 2016a).

2016	PDVSA transferred to Rosneft 49.9% of Citgo, its American subsidiary that manages three refineries and oil pipelines in the United States, as a pledge under a pre-payment contract (Seekingalpha 2016).
2016	Rosneft was announced winner of the tender for the construction of the Tuban refining and petrochemical complex in Indonesia (HydrocarbonTechnology 2016).
2016	Rosneft invited ChemChina to a 40% stake in its Eastern Petrochemical Company (Rosneft 2016c).
2016	Oil production at Vankor began to decline (Reuters 2016).
2017	Igor Sechin missed the court proceedings of the Ulyukaev case for the second time (Mironenko 2017; Reiter 2017; RFERL 2017c).
2017	Rosneft and Bashneft filed a claim with the Moscow Arbitration Court against AFK Sistema, the former owner of Bashneft, for RUB 106.6 billion. The court ultimately ruled that RUB 136.3 billion should be recovered from Sistema (BBC 2017; Polivanov 2017; Solodkov 2017).
2017	Rosneft decided not to build a refinery in Chechnya (RFERL 2017a).
2017	A conflict arose between Rosneft and the Chechen authorities over Chechenneftekhimprom (RFERL 2017b).
2017	QIA and Glencore agreed to sell the lion's share of their 14.16% stake to the largely unknown Chinese company, Clean Energy Finance Corporation (CEFC) (Daily Mail 2017).
2017	Rosneft acquired Kondaneft in its entirety (Rosneft 2017a).
2017	The European Court of Justice ruled that the EU sanctions against Rosneft were legitimate (Rosneft 2017b).
2017	Rosneft drilled a well in the Laptev Sea and announced the discovery of a field there (Rosneft 2017c).
2017	The joint venture between Rosneft and Statoil (now Equinor) began drilling the Domanic formation in Russia's Samara region (Rosneft 2017d).
2017	Rosneft sold 20% in Verkhnechonskneftegaz to Beijing Gas and signed a provisional agreement to supply gas to China (Rosneft 2017e).
2017	Rosneft signed a MoU with BP concerning the sale and purchase of natural gas in Europe (Rosneft 2017f).
2017	Rosneft closed a deal to buy 30% in the licence for the Zohr field in the Mediterranean (Rosneft 2017g).
2017	Rosneft signed a wide-ranging strategic agreement with National Iranian Oil Company (NIOC) (Rosneft 2017h).
2017	Rosneft signed a cooperation agreement and several contracts with Kurdistan (Rosneft 2017i).
2017	Rosneft-Brazil started drilling the first exploration well in the Solimoes project (Rosneft 2017j).
2017	Rosneft sold its stake in the Saras refinery in Italy (Eurasiatx 2017).
2017	Russia became the largest oil supplier to China (Asia Dialogue 2017).
2017	Rosneft and CEFC Energy signed a strategic cooperation agreement during Xi Jinping's official visit to Russia (Rosneft 2017k).
2017	Rosneft in cooperation with partners bought EOL, owner of the Vadimar refinery in India (Reuters 2017).

2017	S&P confirmed its long-term credit rating of Rosneft at BB+ with a 'positive' outlook (Rosneft 2019c).
2018	Moody's upgraded Rosneft's rating to investment level again, with a Baa3 rating and a 'stable' outlook (Rosneft 2019c).
2018	Bashneft, controlled by Rosneft, discovered an oilfield in Iraq (Rosneft 2019a).
2018	Ye Tsianmin, the board chairman of CEFC, was detained on charges of economic crimes, and CEFC's deal with Glencore and QIA collapsed (Reuters 2018).
2019	Rosneft started direct sales of refined products to three refineries in Germany (Rosneft 2019b).

REFERENCES

Asia Dialogue (2017), 'Russia not Saudi Arabia is China's main source of oil', accessed 4 March 2019 at http://theasiadialogue.com/2018/03/28/the-new-king-of-chinas-crude-oil-imports-russia-and-the-competition-for-market-share-in-china/.

BBC (2017), 'Rosneft vs AFK Sistema: Istoriya konflikta', accessed 21 September 2017 at http://www.bbc.co.uk/guides/zwhx8mn.

Caroll, O. (2017), 'Exotic locations, alleged bungs, secret recordings and a basket of sausages: Putin's friend and a no-show in court for Russia's landmark state oil bribery case', accessed 14 November 2017 at https://www.independent.co.uk/news/world/europe/alexei-ulyukayev-bribery-trial-russian-economics-minister-rosneft-igor-sechin-extortion-vladimir-a8051926.html.

China Daily (2007), 'Sinopec, Rosneft deal on Sakhalin project', accessed 4 March 2019 at http://www.chinadaily.com.cn/business/2007-03/30/content_840389.htm.

DailyMail (2017), 'Glencore, Qatar finalize Rosneft deal', accessed 4 March 2019 at https://www.dailymail.co.uk/wires/afp/article-4986094/Glencore-Qatar-finalise-Rosneft-deal.html.

Eurasiatx (2017), 'Rosneft withdraws from Italian Saras', accessed 4 March 2019 at http://eurasiatx.com/rosneft-withdraws-italian-saras/.

Golubkova, K., D. Zhdannikov and S. Jewkes (2016), 'How Russia sold its oil jewel: Without saying who bought it', accessed 21 September 2017 at https://www.reuters.com/article/us-russia-rosneft-privatisation-insight/how-russia-sold-its-oil-jewel-without-saying-who-bought-it-idUSKBN1582OH.

Hydrocarbontechnology (2016), 'Rosneft and Pertamina to develop $13bn Tuban oil refinery in Indonesia', accessed 4 March 2019 at https://www.hydrocarbons-technology.com/news/newsrosneft-pertamina-develop-13bn-tuban-oil-refinery-indonesia-4905914/.

Industryruss (2015), 'Rosneft signs agreements with BP', accessed 4 March 2019 at https://industryruss.wordpress.com/2014/05/27/rosneft-signs-agreements-with-bp/.

Katona, V. (2016), 'What you need to know about Rosneft', accessed 20 September 2017 at https://russia-direct.org/opinion/what-you-need-know-about-rosneft.

Lenta.ru (2012), 'Rosneft', accessed 19 September 2018 at https://lenta.ru/lib/14166182/.

Marchmonthnews (2010), 'Rosneft to build $545m oil refinery in Chechnya', accessed 4 March 2019 at http://marchmontnews.com/Materials-Extraction/Volga/11210-Rosneft-build-545m-oil-refinery-Chechnya.html.

Mironenko, P. (2017), 'The Bell nashel zapisi peregovorov Ulyukaeva i Sechina', accessed 27 November 2017 at https://thebell.io/the-bell-nashel-zapisi-peregovorov-ulyukaeva-i-sechina-v-den-zaderzhaniya-ministra/.

NGFR (2005), 'Rosneft', accessed 21 September 2017 at http://www.ngfr.ru/library.html?rosneft.

Novatek (2005), 'Novatek and Rosneft enter cooperation agreement', accessed 4 March 2019 at http://www.novatek.ru/en/press/releases/archive/index.php?id_4=260&mode_4=all&afrom_4=01.01.1990&from_4=84.

OffshoreEnergy (2016a), 'Pertamina to buy stake in Rosneft's Chayvo field', accessed 4 March 2019 at https://www.offshoreenergytoday.com/pertamina-to-buy-stake-in-rosnefts-chayvo-field/.

OffshoreEnergy (2016b), 'Rosneft: COSL to drill two wells in Okhotsk Sea', accessed 4 March 2019 at https://www.offshoreenergytoday.com/rosneft-cosl-to-drill-two-wells-in-okhotsk-sea/.

Osipov, I. (2013), 'Put Rosnefti: Ot raspada Soyuza do pokupki TNK-BP', accessed 25 February 2017 at http://www.Forbes.ru/novosti-photogallery/236077-put-rosnefti-ot-raspada-soyuza-do-pokupki-tnk-bp?photo=1.

Polivanov, A. (2017), 'Rosneft protiv AFK Sistema – glavnaya istoriya sovremennogo rossiyskogo biznesa. I vot pochemu', accessed 21 September 2017 at https://meduza.io/slides/rosneft-protiv-afk-sistema-glavnaya-istoriya-sovremennogo-rossiyskogo-biznesa-i-vot-pochemu.

Ppnf.ru (2011), 'AO Rosneft', accessed 18 September 2018 at http://ppnf.ru/oao/rosneft/.

Processing (2014), 'Rosneft, PDVSA sign collaboration agreements', accessed 4 March 2019 at https://www.processingmagazine.com/rosneft-pdvsa-sign-collaboration-agreements/.

Reiter, S. (2017), 'Pervyy dopros Ulyukaeva: Nigde i nikogda ya ne treboval u nego nikakuyu vzyatku', accessed 27 November 2017 at https://thebell.io/ulyukaev-pervye-pokazaniya/.

Reuters (2007), 'Rosneft, TNK-BP agree to share disputed oil stake', accessed 4 March 2019 at https://uk.reuters.com/article/rosneft-tnkbp-shares-idUKL2312292920070523.

Reuters (2009), 'China lends Russia $25 billion to get 20 years of oil', accessed 4 March 2019 at https://uk.reuters.com/article/uk-russia-china-oil-sb/china-lends-russia-25-billion-to-get-20-years-of-oil-idUKTRE51G3S620090217.

Reuters (2013), 'Rosneft to double oil flows to China in $270 billion deal', accessed 4 March 2019 at https://www.reuters.com/article/us-rosneft-china/rosneft-to-double-oil-flows-to-china-in-270-billion-deal-idUSBRE95K08820130621.

Reuters (2016), 'Russia's Rosneft says oil output at Vankor field to decline to 21 mln T this year', accessed 4 March 2019 at https://af.reuters.com/article/commoditiesNews/idAFR4N0ZC02O.

Reuters (2017), 'Rosneft seals first Asian refinery deal with Essar Oil purchase', accessed 4 March 2019 at https://www.reuters.com/article/us-india-essar-rosneft/rosneft-seals-first-asian-refinery-deal-with-essar-oil-purchase-idUSKCN1B10PL.

Reuters (2018), 'China's CEFC chairman investigated for suspected economic crimes: Source', accessed 4 March 2019 at https://www.reuters.com/article/us-china-cefc-probe/chinas-cefc-chairman-investigated-for-suspected-economic-crimes-source-idUSKCN1GD3O9.

Revonenko, A. (2017), 'Sechin protiv sistemy. Istorija protivostoyaniya', accessed 21 September 2017 at https://openrussia.org/notes/710925/RFE/RL(2017).

RFERL (2017a), 'Chechnya's Kadyrov, Rosneft again at odds', accessed 4 March 2019 at https://www.rferl.org/a/caucasus-report-chechnya-kadyrov-rosneft-sechin -dispute/28692549.html.

RFERL (2017b), 'Prospects for Chechnya's oil sector remain unclear', accessed 4 March 2019 at https://www.rferl.org/a/caucasus-report-chechnya-oil-sector -prospects-kadyrov/28221846.html.

RFERL (2017c), 'Russian court repeats summons after State Oil Company Chief Sechin skips hearing', accessed 14 November 2017 at https://www.rferl.org/a/ rosneft-sechin-court-summons-extortion-trial-ulyukayev/28850567.html.

Rigzone (2004), 'Rosneft and KNOC team up in Russian Far East', accessed 4 March 2019 at https://www.rigzone.com/news/oil_gas/a/16543/rosneft_and_knoc_team _up_in_russian_far_east/.

Rosneft (2011), 'Rosneft and BP form global and Arctic strategic alliance', accessed 4 March 2019 at https://www.rosneft.com/press/releases/item/114519/.

Rosneft (2012a), 'Rosneft and Sberbank reach agreement in principle on Taas-Yuryakh project', accessed 4 March 2019 at https://www.rosneft.com/press/releases/item/ 114488/.

Rosneft (2012b), 'Rosneft and ITERA Group sign strategic cooperation agreement', accessed 4 March 2019 at https://www.rosneft.com/press/releases/item/114485/.

Rosneft (2012c), 'Rosneft and ITERA Group close deal to create joint venture to produce and sell gas', accessed 4 March 2019 at https://www.rosneft.com/press/ releases/item/114452/.

Rosneft (2013a), 'Rosneft acquires Enel stake in SeverEnergia', accessed 4 March 2019 at https://www.rosneft.com/press/releases/item/114329/.

Rosneft (2013b), 'Rosneft and Novatek agree assets swap', accessed 4 March 2019 at https://www.rosneft.com/press/releases/item/84244/.

Rosneft (2013c), 'Rosneft consolidates 100% of Taas-Yuryakh Neftegazodobycha', accessed 4 March 2019 at https://www.rosneft.com/press/releases/item/24360/.

Rosneft (2013d), 'Rosneft and ExxonMobil expand strategic cooperation', accessed 4 March 2019 at https://www.rosneft.com/press/releases/item/11409/.

Rosneft (2013e), 'Rosneft subsidiary acquires interest in ExxonMobil Gulf of Mexico exploration blocks', accessed 4 March 2019 at https://www.rosneft.com/press/ releases/item/114405/.

Rosneft (2013f), 'Rosneft acquires minority stake in Saras S.p.A.', accessed 4 March 2019 at https://www.rosneft.com/press/releases/item/114399/.

Rosneft (2013g), 'Rosneft acquires ALROSA oil and gas assets', accessed 4 March 2019 at https://www.rosneft.com/press/releases/item/23709/.

Rosneft (2014a), 'Rosneft acquires Orenburg Drilling Company', accessed 4 March 2019 at https://www.rosneft.com/press/releases/item/125322/.

Rosneft (2014b), 'Rosneft signed memorandum of cooperation with Cuba Petroleo', accessed 4 March 2019 at https://www.rosneft.com/press/releases/item/153292/.

Rosneft (2014c), 'Rosneft Brazil and HRT sign final agreements on the Solimoes Project', accessed 4 March 2019 at https://www.rosneft.com/press/releases/item/ 119118/.

Rosneft (2014d), 'Rosneft to acquire Bishkek Oil Company', accessed 4 March 2019 at https://www.rosneft.com/press/releases/item/95189/.

Rosneft (2014e), 'Rosneft increases its stake in the National Oil Consortium to 80% of shares', accessed 4 March 2019 at https://www.rosneft.com/press/releases/item/ 173609/.

Rosneft (2015a), 'Rosneft divest its shares in LLC Polar Lights Company', accessed 4 March 2019 at https://www.rosneft.com/press/releases/item/179543/.

Rosneft (2015b), 'Rosneft and BP complete transaction to sell 20% share of Taas-Yuryakh Neftegazodobycha to BP', accessed 4 March 2019 at https://www.rosneft.com/press/releases/item/178483/.

Rosneft (2015c), 'Rosneft and Sinopec signed a memorandum of understanding on cooperation in gas and petroleum chemicals projects in East Siberia', accessed 4 March 2019 at https://www.rosneft.com/press/releases/item/179545/.

Rosneft (2015d), 'Rosneft acquires a Russian oilfield service company from Trican Well Service Ltd', accessed 4 March 2019 at https://www.rosneft.com/press/releases/item/174453/.

Rosneft (2016a), 'BP and Rosneft create joint venture to develop prospective resources in East and West Siberia', accessed 4 March 2019 at https://www.rosneft.com/press/releases/item/182641/.

Rosneft (2016b), 'Rosneft and BP conclude dissolution of the refining joint venture Ruhr Oel GmbH in Germany', accessed 4 March 2019 at https://www.rosneft.com/press/releases/item/185259/.

Rosneft (2016c), 'Rosneft and ChemChina signed an agreement setting out the framework for further implementation of the FEPCO project', accessed 4 March 2019 at https://www.rosneft.com/press/releases/item/183517/.

Rosneft (2016d), 'Rosneft acquires Targin Oilfield Services Company', accessed 4 March 2019 at https://www.rosneft.com/press/releases/item/185193/.

Rosneft (2017a), 'Rosneft acquires a strategic asset in the new production cluster in Khanty-Mansiysk Autonomous District', accessed 4 March 2019 at https://www.rosneft.com/press/releases/item/186209/.

Rosneft (2017b), 'Rosneft is disappointed by the decision of the EU Court of Justice concerning the estimation of legitimacy of the European sanctions', accessed 4 March 2019 at https://www.rosneft.com/press/releases/item/186035/.

Rosneft (2017c), 'Rosneft discovers hydrocarbon deposits on Eastern Arctic shelf', accessed 4 March 2019 at https://www.rosneft.com/press/releases/item/186997/.

Rosneft (2017d), 'Rosneft and Statoil started pilot drilling as part of the development of Domanik sediments', accessed 4 March 2019 at https://www.rosneft.com/press/releases/item/185549/.

Rosneft (2017e), 'Rosneft and Beijing Gas close the deal for sale and purchase of 20% shares in Verkhnechonskneftegaz', accessed 4 March 2019 at https://www.rosneft.com/press/releases/item/187075/.

Rosneft (2017f), 'BP and Rosneft agree strategic cooperation in gas business', accessed 4 March 2019 at https://www.rosneft.com/press/releases/item/186763/.

Rosneft (2017g), 'Rosneft closes the deal to acquire a 30% stake in Zohr gas field', accessed 4 March 2019 at https://www.rosneft.com/press/releases/item/188045/.

Rosneft (2017h), 'Rosneft and National Iranian Oil Company sign oil and gas strategic cooperation agreement', accessed 4 March 2019 at https://www.rosneft.com/press/releases/item/188381/.

Rosneft (2017i), 'Rosneft and the government of the Kurdish Autonomous Region of Iraq agree on cooperation at five production blocks', accessed 4 March 2019 at https://www.rosneft.com/press/releases/item/188125/.

Rosneft (2017j), 'Rosneft starts drilling the first exploration well at the Solimoes Project', accessed 4 March 2019 at https://www.rosneft.com/press/releases/item/185729/.

Rosneft (2017k), 'Rosneft and CEFC deepen strategic cooperation', accessed 4 March 2019 at https://www.rosneft.com/press/releases/item/188599/.

Rosneft (2017l), 'History of Rosneft', accessed 17 September 2017 at https://www.rosneft.com/about/History/.

Rosneft (2019a), 'Rosneft discovers new oil field in Iraq', accessed 8 February 2019 at https://www.rosneft.com/press/news/item/191233/.

Rosneft (2019b), 'Rosneft Deutschland successfully started direct marketing operations', accessed 8 February 2019 at https://www.rosneft.com/press/news/item/191527/.

Rosneft (2019c), 'Ratings', accessed 4 March 2019 at https://www.rosneft.com/Investors/Instrumenti_dlja_investora/Rejtingi/.

Seekingalpha (2016), 'Rosneft could take control of 49.9% of Citgo if PDVSA defaults', accessed 4 March 2019 at https://seekingalpha.com/news/3251534-rosneft-take-control-49_9-percent-citgo-pdvsa-defaults.

Solodkov, A. (2017), 'Protiv sistemy: Khronika protsessa po krupneyshemu isku Rosnefti', accessed 21 September 2017 at https://www.rbc.ru/business/07/12/2017/5952376e9a7947d0bb9e5308.

Sputniknews (2005), 'Timeline: Rosneft company history', accessed 19 September 2018 at https://sputniknews.com/russia/200511244219961 5/.

TASS (2016), 'Indian companies invest $5 bln in Vankorneft and TAAS-Yuryakh Neftegazodobycha – ONGC', accessed 4 March 2019 at http://tass.com/economy/905972.

Wikipedia (2018), 'Rosneft', accessed 19 September 2018 at https://en.wikipedia.org/wiki/Rosneft.

Appendix 2: Corporate history of LUKOIL

1960	The Shaimskoye oilfield was discovered in 1960. The nearby settlement of Urai became a hub for oil workers and was declared a town in 1965 (LUKOIL 2018a).
1962	The history of Langepas oil company started when the Lokosovskoye oilfield was discovered in 1959 (LUKOIL 2018a).
1972	The town of Kogalym was founded in 1975 after the discovery of oilfields in the vicinity.
1991	On 25 November 1991, the government created LangepasUraiKogalym (LUKOIL), consolidating the oil-producing enterprises of Langepas, Urai and Kogalym as well as several refineries, including those in Perm and Volgograd (LUKOIL 2018b).
1993	The Council of Ministers incorporated the public joint stock company (JSC) LUKOIL. Vagit Alekperov was appointed the company's President, CEO and chairman of the board of directors. A privatization programme was approved, and the first issue of LUKOIL shares was registered (LUKOIL 2018b).
1993	On the initiative of the company's management, the LUKOIL Charity Fund was established, one of the first corporate charity funds in post-Soviet Russia (LUKOIL 2018b).
1994	LUKOIL acquired a 10% stake in the project for the development of Azeri-Chirag-Gyuneshli, the largest oilfield in the Azerbaijani sector of the Caspian Sea (LUKOIL 2018b).
1994	The trade unions of the major oil production, refining and petroleum product distributors of LUKOIL were merged into one trade union, the 'Inter-Regional Trade Union Organization', with a total of about 90 000 members (LUKOIL 2018b).
1995	A 5% state stake in LUKOIL was sold at a loans-for-shares auction, and 16.07% was offered in an investment tender. Shares were bought by LUKOIL itself at the loans-for-shares auction and by NIKoil in the tender (Salomon Brothers 1996).
1995	Stakes in nine oil-producing, marketing and service enterprises in Western Siberia, and the Volga-Urals regions were added to the company's authorized capital (LUKOIL 2018b).
1995	ARCO acquired a 6.3% stake in LUKOIL and became a strategic partner (LUKOIL 2018b).
1995	LUKOIL joined the Kumkol project in Kazakhstan and the Meleya project in Egypt (LUKOIL 2018b).
1995	LUKOIL issued first-level American depository receipts (ADRs) (NGFR 2010).
1996	LUKOIL acquired a 5% stake in the Shakh-Deniz international gas project in the Azerbaijani sector of the Caspian Sea. In 2004, the company increased its stake in the project to 10% (LUKOIL 2018b).
1996	LUKOIL and ENI established the LukAgip joint venture (NGFR 2010).
1996	LUKOIL and ARCO established LukArco joint venture (New York Times 1996).

1996   LUKOIL began to establish its own tanker fleet (NGFR 2010).

1997   LUKOIL acquired a 15% stake in the project for the development of Karachaganak gas field and a 5% stake in Tengiz project, both in Kazakhstan (LUKOIL 2018b).

1997   LUKOIL Racing Team had become one of the leaders in Russian motor racing (LUKOIL 2018b).

1997   LUKOIL bought a controlling interest in Arkhangelskgeoldobycha and, in 2003, increased its stake to 99.7% (NGFR 2010).

1998   LUKOIL acquired a controlling stake in Petrotel refinery located in Ploieşti, Romania (LUKOIL 2018b).

1999   LUKOIL acquired KomiTEK, becoming a dominant player in Timan-Pechora (NGFR 2010).

1999   LUKOIL acquired controlling stakes in Odessa refinery (Ukraine), Stavropolpolymer (later renamed Stavrolen), Saratovorgsintez petrochemical plant and became the key shareholder of the Bulgarian Neftokhim petrochemical enterprise in Burgas (LUKOIL 2018b).

2000   LUKOIL started developing the Kravtsovskoye field in the Baltic Sea (NGFR 2010).

2000   Following the acquisition of ARCO, BP gained a 7% stake in LUKOIL. In early 2001, BP announced its intention to sell the stake. In January 2003, BP started converting the bonds into the company's shares, thus withdrawing from the authorized capital of LUKOIL (LUKOIL 2018b).

2000   LUKOIL entered the US petroleum products retail market through the acquisition of Getty Petroleum Marketing, which operated 1260 service stations in 13 north-eastern states.

2001   LUKOIL acquired Yamalgazneftedobycha, a holder of subsoil use licences in the Bolshekhetskaya Depression in Yamal-Nenets Autonomous District (LUKOIL 2018b).

2001   LUKOIL acquired the state's stake in NORSI-Oil, the owner of Nizhegorodnefteorgsyntez refinery (NGFR 2010).

2001   LUKOIL bought the Lokosovskiy gas-processing complex from SIBUR (NGFR 2010).

2002   LUKOIL settled a conflict with tax authorities and voluntarily gave up the use of the so-called Baikonur scheme and paid taxes (Vedomosti 2005).

2002   LUKOIL began to divest itself of non-core assets (Forbes 2007).

2002   Sergey Kukura, Chief Financial Officer of LUKOIL, was kidnapped by five masked men (Telegraph 2002).

2002   LUKOIL signed a contract with Colombia's NOC Ecopetrol for the joint oil exploration and production at Condor block (LUKOIL 2018b).

2003   LUKOIL purchased a controlling stake in Serbia's Beopetrol, which controls about 20% of the Serbian retail fuel market (LUKOIL 2018b).

2003   President Vladimir Putin participated in the official opening ceremony of LUKOIL's service station in New York (LUKOIL 2018b).

2003   LUKOIL Overseas Egypt signed a concessionary agreement with Egypt to develop the North-East Geisum and West Geisum blocks (NGFR 2010).

2003   LUKOIL sold its stake in Azeri-Chirag-Guneshli to Inpex (Neft i kapital 2003).

2004   LUKOIL-Western Siberia began the development of the Nakhodkinskoye gas field (NGFR 2010).

2004	ConocoPhillips submitted a winning bid to purchase a 7.59% stake in LUKOIL previously owned by the state. In 2007, ConocoPhillips increased its stake in LUKOIL to 20% (LUKOIL 2018b).
2004	LUKOIL purchased a 50% stake in the production-sharing agreement (PSA) for Tyub-Karagan offshore block in the Kazakh sector of the Caspian Sea (LUKOIL 2018b).
2004	LUKOIL and Saudi Aramco signed a 40-year contract for the exploration and development of gas and gas-condensate fields at Block A in Saudi Arabia (LUKOIL 2018b).
2004	LUKOIL and the Uzbekneftegaz signed a PSA for Kandym-Khauzak-Shady project giving LUKOIL a 90% stake in the project (LUKOIL 2018b).
2004	LUKOIL started publishing biannual sustainable development reports. From 2018 they were published annually (LUKOIL 2018b).
2004	A conflict arose between LUKOIL and the Azeri tax authorities. The Azeris accused LUKOIL of tax evasion, which the company denied (REGNUM 2004).
2004	LUKOIL commissioned the first stage of the oil terminal in Vysotsk (Leningrad region) (NGFR 2010).
2005	LUKOIL discovered a major multi-reservoir oil-gas-condensate field in the Severny licence area in the northern part of the Caspian Sea (LUKOIL 2018b).
2005	LUKOIL acquired the Finnish petroleum product distributors Teboil and Suomen Petrooli (LUKOIL 2018b).
2005	LUKOIL and PDVSA signed an agreement on the development of the Junin-3 block in Venezuela (RIA 2011a).
2005	LUKOIL acquired Nelson Resources Limited, which held stakes in four production projects in Western Kazakhstan and options on two exploration blocks in the Kazakh sector of the Caspian Sea (LUKOIL 2018b).
2005	LUKOIL-Western Siberia bought 66% of Geoilbent from Novatek and in 2007 bought another 34% of Geoilbent from Russneft (NGFR 2010).
2005	LUKOIL and ConocoPhillips established the Naryanmarneftegaz joint venture (Forbes 2012).
2005	LUKOIL discovered the major oilfield Filanovskogo in the Caspian Sea (NGFR 2010).
2005	LUKOIL signed a strategic partnership agreement with Gazprom (RIA 2005).
2005	LUKOIL was Russia's first oil company to commence large-scale production of Euro-4 diesel fuel (LUKOIL 2018b).
2006	LUKOIL acquired a 63% stake in the PSA on exploration, development and production of hydrocarbons at a deep-water block in Cote d'Ivoire (LUKOIL 2018b).
2006	LUKOIL bought producing assets in Khanty-Mansi Autonomous District from Marathon Oil (NGFR 2010).
2007	LUKOIL agreed with Vanco Energy, an American oil company, to buy a 56.66% stake in three projects for exploration of prospective offshore blocks in the Gulf of Guinea in Western Africa (LUKOIL 2018b).
2007	The small American company Green Oil accused LUKOIL of inflating the price of its oil products in the United States (Skandaly.ru 2007).

2008	Naryanmarneftegaz began oil production from the Yuzhno-Khylchuyusk field (Kommersant 2012a; RIA 2005).
2008	LUKOIL launched the Varandei oil terminal in the Barents Sea (LUKOIL 2019b).
2008	LUKOIL (49%) and Italy's ERG (51%) established a joint venture to operate ISAB, a major oil-refining facility in Sicily (LUKOIL 2018b).
2008	LUKOIL acquired Akpet, a Turkish company operating 693 service stations under dealer agreements, making up about 5% of the Turkish market (LUKOIL 2018b).
2008	LUKOIL joined the National Oil Consortium in Venezuela with a 20% stake (Kommersant 2013).
2009	LUKOIL finalized a deal to acquire a 45% stake in TOTAL Raffinaderij Nederland from Total (LUKOIL 2018b).
2009	LUKOIL acquired a 46% stake in the LukArco joint venture from BP. As a result, the company received a 5% stake in the Tengizchevroil joint venture, which was developing the Tengiz and Korolevskoye fields in Kazakhstan (LUKOIL 2018b).
2010	LUKOIL commissioned the Korchagin field in the Caspian Sea (LUKOIL 2019c).
2010	LUKOIL won a bid for the development of the West Qurna-2 oilfield in Iraq (LUKOIL 2018b).
2010	A high-ranking LUKOIL employee was involved in a fatal car crash but was found not guilty of causing the crash (Radio Liberty 2010).
2010	A consortium consisting of LUKOIL, American Vanco and Ghana National Petroleum Corporation discovered significant hydrocarbon reserves in the Dzata structure on Ghana's continental shelf (LUKOIL 2018b).
2011	ConocoPhillips sold its stake in LUKOIL (LUKOIL 2018b).
2011	LUKOIL purchased a 50% stake in the Vietnamese Hanoi Trough from Quad Energy (LUKOIL 2018b).
2011	LUKOIL and Italian ERG Renew signed an agreement for the establishment of the LUKERG Renew joint venture to develop renewable energy (LUKOIL 2018b).
2011	LUKOIL acquired from Oranto Petroleum a 49% stake in the exploration and production contract for a deep-water block outside Sierra Leone (LUKOIL 2018b).
2011	Jointly with American Vanco and PETROCI, LUKOIL made a discovery on Cote d'Ivoire's continental shelf (LUKOIL 2018b).
2011	LUKOIL sold Getty Petroleum Marketing (RIA 2011b).
2011	As part of a consortium, LUKOIL received the licence for the Trident and Rapsodia blocks in Romania (RIA 2015).
2011	Production at the Yuzhno-Khylchuyusk field began to decline and its reserves were downgraded (Forbes 2012).
2011	LUKOIL joined forces with Bashneft in the Bashneft Polyus joint venture, the licence holder for the Trebs and Titov fields in Timan-Pechora (Kommersant 2012a).
2011	LUKOIL terminated its activities on the Junin-3 block in Venezuela (RIA 2011a).
2011	LUKOIL production declined for the first time (Kommersant 2012b).
2011	The Bulgarian authorities suspected LUKOIL of fraudulent exports and revoked the licence of Neftokhim Burgas (Kommersant 2011).

2012	Agreements were signed with Verolma Group to acquire 46 petrol stations in the Netherlands, 13 petrol stations in Belgium and eight petrol stations in Belgium from NGM Group (LUKOIL 2018b).
2012	LUKOIL sold its stake in the Condor block in Colombia (Kommersant 2012c).
2012	LUKOIL and Inpex won the licence for Block 10 in Iraq (Kommersant 2012d).
2012	LUKOIL launched the production of a new generation of engine oils at the Petrotel-LUKOIL refinery in Romania (LUKOIL 2018b).
2012	LUKOIL bought 30% of Naryanmarneftegaz from ConocoPhillips, becoming its sole owner (Forbes 2012).
2012	LUKOIL won the licence for the Imilorskoye group of fields (Kommersant 2012a).
2013	LUKOIL agreed to pay USD 93 million to settle a lawsuit arising from the bankruptcy of Getty Petroleum Marketing (World Oil 2013).
2013	LUKOIL sold its Odessa refinery (Neft I kapital 2013).
2013	LUKOIL joined two projects in the Norwegian sector of the Barents Sea (LUKOIL 2018b).
2013	The joint venture of LUKOIL and Italian ERG Renew, LUKERG Renew, acquired Land Power of Romania (LUKOIL 2018b).
2013	LUKOIL acquired a 65% stake in an offshore block in the waters of Cote d'Ivoire in the Gulf of Guinea (LUKOIL 2018b).
2013	LUKOIL and OMV Refining & Marketing signed a contract for the acquisition of OMV's lubricants plant outside Vienna (LUKOIL 2018b).
2013	An accident at LUKOIL's refinery in Bulgaria seriously injured four people (Sofia News Agency 2013).
2013	LUKOIL acquired the remaining 20% stake in the joint venture created to operate the ISAB refinery in Sicily from Italian ERG (LUKOIL 2018b).
2013	LUKOIL sold its stake in the National Oil Consortium to Rosneft (Kommersant 2013).
2014	LUKOIL signed a deal on oil exploration and production with Mexico's NOC Pemex (The Borgen Project 2014).
2014	In cooperation with Saudi Aramco, LUKOIL drilled two evaluation wells in the Mushaib tight gas field in the Empty Quarter (Reuters 2014).
2014	The Romanian government accused LUKOIL of tax evasion and money laundering (Vkrizis.ru 2014).
2014	A new home arena of Spartak Moscow football club was opened with LUKOIL as the general sponsor (LUKOIL 2018b).
2014	Three members of staff in connection with LUKOIL's offshore operations in Ghana were killed in a helicopter crash (Offshore Energy Today 2014).
2014	LUKOIL was added to the list of companies affected by the US sanctions against Russia (Forbes 2014).
2014	LUKOIL and Gazprom extended their partnership agreement up to 2024 (LUKOIL 2018c).
2014	LUKOIL formed a joint venture with Total to work on the Bazhenov play in Western Siberia (LUKOIL 2018b).

2015	LUKOIL and WWF signed a cooperation agreement (LUKOIL 2018b).
2015	LUKOIL, PanAtlantic and Romgaz discovered a major offshore field in the Lira marine structure on the Romanian continental shelf (LUKOIL 2018b).
2016	LUKOIL commissioned Filanovskogo field in the Caspian Sea (LUKOIL 2019d).
2016	Azat Shamsuarov, son of LUKOIL's vice president, was involved in a car chase with the police for the third time and sentenced to 15 days in prison (Vkrizis.ru 2016).
2016	*Forbes* wrote that Heesen Yachts, a well-known Dutch producer of superyachts, belongs to the President of LUKOIL, Vagit Alekperov (Forbes 2015).
2016	The vice president of LUKOIL notified Bashneft that LUKOIL intended to acquire 25% of Bashneft (Skandaly.ru 2016).
2016	LUKOIL withdrew from the Rapsodia project in Romania (Neft i kapital 2017).
2016	LUKOIL withdrew from Block A in Saudi Arabia (RBC 2016a).
2016	LUKOIL sold its fuel stations in Latvia, Lithuania and Poland (RBC 2016b).
2017	LUKOIL was involved in a tax evasion case in Romania (Romania Insider 2017).
2017	Four people died in a fire at the LUKOIL plant in Kstov, Nizhegorodskoy Oblast (Ruptly 2017).
2018	LUKOIL reported 20 million tonnes of oil produced from North Caspian fields (World Oil 2018).
2018	LUKOIL rebuffed the interest of Rosneft in the Trebs and Titov project (Skandaly.ru 2018).
2019	LUKOIL started drilling new production wells at the West Qurna-2 field in Iraq (LUKOIL 2019a).

REFERENCES

Forbes (2007), 'LUKOYL pro zapas', accessed 1 March 2019 at https://www.forbes.ru/Forbes/issue/2007-06/12377-lukoil-pro-zapas.

Forbes (2012), 'LUKOYL vykupil 30% "Naryanmarneftegaza" u ConocoPhillips', accessed 1 March 2019 at https://www.forbes.ru/news/98270-lukoil-vykupil-30-naryanmarneftegaza-u-conocophillips.

Forbes (2014), 'LUKOIL pod sanktsiyami: Pochemu v spisok SSHA popala chastnaya kompaniya', accessed 1 March 2019 at https://www.forbes.ru/kompanii/resursy/267605-lukoil-pod-sanktsiyami-pochemu-v-spisok-ssha-popala-chastnaya-kompaniya.

Forbes (2015), 'Verfi Alekperova: Vladelets LUKOYLa vkladyvayet v stroitelstvo elitnykh yakht', accessed 8 February 2019 at http://www.forbes.ru/milliardery/288331-verfi-alekperova-vladelets-lukoila-vkladyvaet-v-stroitelstvo-elitnykh-yakht.

Kommersant (2011), 'LUKOYL otbil litsenziyu', accessed 1 March 2019 at https://www.kommersant.ru/doc/1689728.

Kommersant (2012a), 'LUKOYL zabral u Rossii samoye dorogoye', accessed 1 March 2019 at https://www.kommersant.ru/doc/2098523.

Kommersant (2012b), 'LUKOYL rasplatilsya za oshibki geologov', accessed 1 March 2019 at https://www.kommersant.ru/doc/1884165.

Kommersant (2012c), 'LUKOYL vypustil Kondora', accessed 1 March 2019 at https://www.kommersant.ru/doc/2088810.

Kommersant (2012d), 'Neftyanoye peremiriye', accessed 1 March 2019 at https://www.kommersant.ru/doc/2041743.

Kommersant (2013), 'Venesuelu ostavyat goskompaniyam', accessed 1 March 2019 at https://www.kommersant.ru/doc/2309815.

LUKOIL (2018a), 'Predpriyatiye', accessed 8 February 2019 at http://zs.lukoil.ru/ru/About/Structure/Langepasneftegaz.

LUKOIL (2018b), 'History', accessed 8 February 2019 at http://www.lukoil.com/Company/history/History1991.

LUKOIL (2018c), 'LUKOIL i Gazprom zaklyuchili soglasheniye o namereniyakh po realizatsii proyekta razrabotki dvukh mestorozhdeniy v nao', accessed 1 March 2019 at http://www.lukoil.ru/PressCenter/Pressreleases/Pressrelease?rid=220909.

LUKOIL (2019a), 'LUKOIL begins drilling new production wells at West Qurna-2', accessed 8 February 2019 at http://www.lukoil.com/PressCenter/Pressreleases/Pressrelease?rid=326315.

LUKOIL (2019b), 'Varandeyskiy terminal', accessed 1 March 2019 at http://trans.lukoil.ru/ru/About/Structure/VarandeyTerminal.

LUKOIL (2019c), 'Mestorozhdeniye im. Yu. Korchagina', accessed 1 March 2019 at http://www.lukoil.ru/Business/Upstream/KeyProjects/KorchaginField.

LUKOIL (2019d), 'Mestorozhdeniye im. V. Filanovskogo', accessed 1 March 2019 at http://www.lukoil.ru/Business/Upstream/KeyProjects/Filanovskyfield.

Neft i kapital (2003), 'LUKOYL zavershayet prodazhu svoyey doli v proyekte Azeri-Chirag-Gyuneshli INPEKSu', accessed 1 March 2019 at https://oilcapital.ru/news/markets/28-04-2003/lukoyl-zavershaet-prodazhu-svoey-doli-v-proekte-azeri-chirag-gyuneshli-inpeksu.

Neft i kapital (2013), 'LUKOYL prodayet Odesskiy NPZ ukrainskomu VETEKu', accessed 1 March 2019 at https://oilcapital.ru/news/companies/05-03-2013/lukoyl-prodaet-odesskiy-npz-ukrainskomu-veteku.

New York Times (1996), 'ARCO looks to Russia to fill its vast appetite for oil reserves', accessed 1 March 2019 at https://www.nytimes.com/1996/09/24/business/arco-looks-to-russia-to-fill-its-vast-appetite-for-oil-reserves.html.

NGFR (2010), 'LUKOIL', accessed 1 March 2019 at http://www.ngfr.ru/library.html?lukoil.

Offshore Energy Today (2014), 'LUKOIL confirms helicopter crash offshore Ghana. Three dead', accessed 9 February 2019 at https://www.offshoreenergytoday.com/lukoil-confirms-helicopter-crash-offshore-ghana-three-dead/.

Radio Liberty (2010), 'LUKOIL official found not guilty in deadly car crash', accessed 9 February 2019 at https://www.rferl.org/a/lukoil_Official_Found_Not_Guilty_In_Deadly_Car_Crash/2151590.html.

RBC (2016a), 'LUKOYL zayavil ob ukhode iz Saudovskoy Aravii', accessed 1 March 2019 at https://www.rbc.ru/rbcfreenews/5762802a9a794724eff78999.

RBC (2016b), 'LUKOYL dogovorilsya o prodazhe zapravok v Pribaltike i Polshe', accessed 1 March 2019 at https://www.rbc.ru/business/05/02/2016/56b497e69a7947ab333a1b5c.

REGNUM (2004), 'Konflikt mezhdu "LUKOYL" i Ministerstvom nalogov Azerbaydzhana prodolzhayetsya', accessed 8 February 2019 at https://regnum.ru/news/220131.html.

Reuters (2014), 'Russia's LUKOIL to drill for tight gas in Saudi desert', accessed 3 October 2017 at https://www.reuters.com/article/saudi-lukoil-gas/russias-lukoil-to-drill-for-tight-gas-in-saudi-desert-idUSL6N0O034G20140515.

RIA (2005), 'Gazprom i LUKOYL zaklyuchili soglasheniye o strategicheskom partnerstve', accessed 1 March 2019 at https://ria.ru/20050329/39587682.html.

RIA (2011a), 'LUKOYL prekratil raboty po proyektu "Khunin-3" v Venesuele', accessed 1 March 2019 at https://ria.ru/20111031/476468540.html.

RIA (2011b), 'Amerikanskaya "dochka" LUKOYLa prodala podrazdeleniye Getty Petroleum', accessed 1 March 2019 at https://ria.ru/20110303/341472213.html.

RIA (2015), 'LUKOYL vlozhit do $300 mln v 2 skvazhiny na shel'fe Rumynii', accessed 1 March 2019 at https://ria.ru/20150402/1056083185.html.

Romania Insider (2017), 'LUKOIL involved in tax evasion case in Romania', accessed 9 February 2019 at https://www.romania-insider.com/lukoil-tax-evasion-romania/.

Ruptly (2017), 'Russia: Four killed in explosion at LUKOIL plant in Nizhny Novgorod – reports', accessed 8 February 2019 at https://www.youtube.com/watch?v=JNYE2kAKgP8.

Salomon Brothers (1996), *Russian Oil Vertically Integrated Companies*, Volume 3 of Report on Russian Oil, March, London: Salomon Brothers, pp. 30–1.

Skandaly.ru (2007), 'LUKOIL sgovorilsya na $ 25 mlrd?', accessed 8 February 2019 at https://scandaly.ru/2007/08/06/lukojl-sgovorilsya-na-25-mlrd/.

Skandaly.ru (2016), 'Kontsy v LUKOYL?', accessed 8 February 2019 at http://scandaly.ru/2016/04/20/kontsyi-v-lukoyl/.

Skandaly.ru (2018), 'Vagit Alekperov vstupil v boy s Igorem Sechinym', accessed 8 February 2019 at https://scandaly.ru/2018/10/02/vagit-alekperov-vstupil-v-boy-s-igorem-sechinyim/.

Sofia News Agency (2013), 'Accident at Bulgaria's LUKOIL refinery leaves 4 seriously injured', accessed 9 February 2019 at https://www.novinite.com/articles/156403/Accident+at+Bulgaria%27s+Lukoil+Refinery+Leaves+4+Seriously+Injured.

Telegraph (2002), 'LUKOIL offers $1m reward to trace kidnapped executive', accessed 14 October 2017 at http://www.telegraph.co.uk/finance/2773437/Lukoil-offers-1m-reward-to-trace-kidnapped-executive.html.

The Borgen Project (2014), 'LUKOIL and PEMEX sign petroleum agreement', accessed 3 October 2017 at https://borgenproject.org/lukoil-pemex-sign-petroleum-agreement/.

Vedomosti (2005), 'LUKOYL otlozhil na nalogi $ 163 mln', accessed 1 March 2019 at https://www.vedomosti.ru/newspaper/articles/2005/09/21/lukojl-otlozhil-na-nalogi-163-mln.

Vkrizis.ru (2014), 'Rumyniya zapodozrila "LUKOYL" v uklonenii ot uplaty nalogov i otmyvanii deneg', accessed 3 October 2017 at http://vkrizis.ru/energetika/rumyiniya-zapodozrila-lukoyl-v-uklonenii-ot-uplatyi-nalogov-i-otmyivanii-deneg/.

Vkrizis.ru (2016), 'Protiv syna vitse-prezidenta "LUKOYLa" v tretiy raz vozbudili ugolovnoye delo', accessed 3 October 2017 at http://vkrizis.ru/obschestvo/protiv-syina-vitse-prezidenta-lukoyla-v-tretiy-raz-vozbudili-ugolovnoe-delo/.

World Oil (2013), 'LUKOIL to pay $93 million to settle Getty petroleum suit', accessed 3 October 2017 at https://www.worldoil.com/news/2013/7/18/lukoil-to-pay-93-million-to-settle-getty-petroleum-suit.

World Oil (2018), 'LUKOIL reports 20 million tonnes of oil produced from North Caspian fields', accessed 8 February 2019 at https://www.worldoil.com/news/2018/12/21/lukoil-reports-20-million-tonnes-of-oil-produced-from-north-caspian-fields.

Appendix 3: Corporate history of Gazprom Neft

1973   The Noyabrsk oilfield was discovered (Live Journal 2017).

1975   The development of the Noyabrsk field and the town of Noyabrsk commenced (Live Journal 2017).

1981   Noyabrskneftegaz production association was established within Glavtummenneftegaz (Live Journal, 2017).

1995   Siberian Oil Company (Sibneft) was created by a presidential decree based on Noyabrskneftegaz, Noyabrskgeophisica, Omsk refinery and Omsknefteprodukt. Victor Gorodilov, Head of Noyabrskneftegaz, was appointed President of the company (Gazprom 2019).

1995   A loans-for-shares auction was held for a 51% stake in Sibneft. The winner was Neftyanaya Finansovaya Kompaniya (NFK) (Sibneft 2019).

1995   The development of the Sugmutskoye field began (Sibneft 2019).

1996   Omsk refinery became a subsidiary of Sibneft (Sibneft 2019).

1996   On 20 September, a tender was held for a 19% stake in Sibneft. The winner was ZAO Sins. On 24 October, another 15% stake in Sibneft was sold. The winner was Refine-Oil (Sibneft 2019).

1997   The Heads of the Sibneft and YUKOS signed a memorandum on merging the two companies. The resulting merged company was to be called Yuksi and would have been the largest oil company in Russia and the third or fourth largest in the world (Sibneft 2019).

1997   In March 1997, Sibneft acquired 47.02% of the East-Siberian Oil and Gas Company with estimated reserves of 700 million tonnes of oil (Sibneft 2019).

1997   Andrei Blokh was appointed President of Sibneft replacing Victor Gorodilov (Sibneft 2019).

1997   Sibneft placed three-year Eurobonds worth USD 150 million (Sibneft 2019).

1998   YUKOS and Sibneft suspended the merger process (Sibneft 2019).

1998   Andrey Bloch resigned from the position of the President of Sibneft and was replaced by Evgeniy Shvidler (Sibneft 2019).

1998   The fall in oil prices and the Russian financial crisis forced Sibneft to cut costs, mainly by curtailing the Noyabrskneftegaz investment programme (Sibneft 2019).

1999   In October 1999, Sibneft sold a controlling stake in the East-Siberian Oil and Gas Company to Rosneftegazstroy. Later, YUKOS became the owner of this stake (Sibneft 2019).

1999   Sibneft's board of directors established a committee to work with minority shareholders (Sibneft 2019).

2000   YUKOS sold Sibneft a 38% stake in Orenburgneft, the main oil-producing subsidiary of ONACO, for USD 430 million (Sibneft 2019).

2000   Sibneft won the tender to develop the south-western part of the Krapivinskoye field located in the Omsk region (Sibneft 2019).

2000   Sibneft produced the first oil from the Yarainerskoye field located in Yamal-Nenets Autonomous District (Sibneft 2019).

2000   Sibneft and Sibir Energy established a joint venture Sibneft-Yugra (Sibneft 2019).

2000   Sibneft acquired a 27% stake in Stavropolneftegaz (Rosneft owned a 38% stake in the company) (Sibneft 2019).

2000   Sibneft acquired controlling stakes in Sverdlovsknefteprodukt and the Yekaterinburg Oil Products Company (Sibneft 2019).

2000   Roman Abramovich was elected Governor of Chukotka (RIA 2008).

2001   Sibneft acquired a 35% stake in the Moscow refinery and a 14.95% stake in Mosnefteproduct from LUKOIL. After that, a lengthy conflict arose between Sibneft, Moscow city authorities and Sibir Energy concerning control over the refinery (Sibneft 2019).

2001   Sibneft and the Chukotka Trading Company established the Sibneft-Chukotka enterprise on a 50:50 basis (Sibneft 2019).

2001   Sibneft acquired 78.4% of the voting shares of Tyumennefteprodukt from TNK with 80 petrol stations and 22 tank farms (Sibneft 2019).

2001   A conflict broke out between NIKoil, an investment banking group, and Sibneft (Kommersant 2001).

2002   Sibneft acquired oil production assets in the Tomsk and Omsk regions and fuel stations in Sverdlovsk and Tyumen Oblasts, Krasnodar Krai, Saint Petersburg and Moscow (Gazprom 2019).

2002   Sibneft commissioned a new oil pipeline linking the south-western part of the Krapivinskoye field with the Transneft trunk system (Sibneft 2019).

2002   On 6 December, Sibneft bought out 10.83% of Slavneft's shares from Belarus. On 18 December, an auction was held to sell a 74.95% stake in Slavneft, which belonged to the Russian government. The winner of the auction was Investoil, created by Sibneft and TNK on a 50:50 basis. Another 12.98% of Slavneft's shares were owned by a trust company, also controlled by Sibneft and TNK. As a result, Sibneft and TNK jointly came to control about 99% of Slavneft (Sibneft 2019).

2002   In April 2002, Sibneft acquired a 67% stake in Meretoyakhaneftegaz from a group of Russian and foreign investors (Sibneft 2019).

2002   Sibneft established a new subsidiary, Sibneft-Krasnoyarsknefteprodukt (Sibneft 2019).

2003   Sibneft established the 100% subsidiary Sibneft Saint Petersburg to develop a sales network in Saint Petersburg (Sibneft 2019).

2003   SIBUR and Sibneft signed a memorandum on the establishment of the joint venture Noyabrsk Gas Energy Company based on Muravlenkovskiy GPP (Sibneft 2019).

2003   An extraordinary meeting of shareholders of Sibneft decided to merge the independent trader UNICAR into the company (Sibneft 2019).

2003   Sibneft and TNK reached a preliminary agreement on the division of Slavneft's assets (Sibneft 2019).

2003   YUKOS and the main shareholders of Sibneft (Millhouse Capital) again reached an agreement, in principle, to merge the companies. The combined company was now to be named YukosSibneft. It would have been Russia's largest in terms of oil production and reserves and the world's second-largest private oil producer. YUKOS had obtained some 92% of Sibneft's shares in return for USD 3 billion and 26% of its own shares (Sibneft 2019).

2003   Sibneft's effective income tax rate in the first half of the year was 5%.

2004   Sibir Energy discovered that its share in Sibneft-Yugra was diluted from 50% to 1%. A new conflict between Sibneft and Sibir Energy ensued (Sibneft 2019).

2004   In February, the Ministry of Taxes and Duties made fiscal claims worth USD 1 billion against Sibneft for 2000–01. The company reached a compromise and paid some USD 300 million to the state (Neftegaz 2005).

2004   In September, YUKOS initiated a lawsuit against Millhouse in the London International Arbitration Court.

2004   In October, YUKOS returned 57.5% of Sibneft's shares to Millhouse Capital (Sibneft 2019).

2005   In July, Millhouse Capital obtained another 14.5% of Sibneft from YUKOS (Sibneft 2019).

2005   Sibneft-Khantos was established, including the Priobskiy, Palyanovskiy, Zimniy and Salym sites (Sibneft 2019).

2005   In 2005, Sibneft bought 75% of TNK-Sakhalin (licence holder for the Lopukhov block) from TNK-BP (Sibneft 2019).

2005   Gazprom and Millhouse Capital closed the deal for the purchase of a 72.7% stake in Sibneft for USD 13 billion. Earlier, Gazprom had bought 3% of Sibneft's stock from Gazprombank. Thus, Gazprom gained control over a 75.7% stake in Sibneft. On 23 December, an extraordinary meeting of Sibneft shareholders elected a new president for the company, Alexander Ryazanov, the deputy chairman of the board of Gazprom (Sibneft 2019).

2005   Roman Abramovich was elected Governor of Chukotka for a second term (RIA 2008).

2006   An extraordinary meeting of shareholders of Sibneft changed the name of the company to Gazprom Neft and its address from Omsk to Saint Petersburg (Sibneft 2019).

2006   Alexander Dyukov, the former President of SIBUR, was appointed President of Gazprom Neft, replacing Alexander Ryazanov (Gazprom Neft 2019).

2006   Sibneft bought 14 multifunctional petrol stations in the Moscow and Tver regions from RK-Gazsetservice (Sibneft 2019).

2006   Gazprom Neft entered the Central Asian retail market with a dedicated subsidiary, Gazprom Neft Asia, selling petroleum products in Kazakhstan, Kyrgyzstan and Tajikistan (Gazprom 2019).

2007   Gazprom Neft acquired a 50% interest in Tomskneft (Gazprom 2019).

2007   Subsidiaries were created within Sibneft for separate business divisions, including Gazpromneft Marine Bunker, Gazpromneft Lubricants and Gazpromneft Aero (Gazprom 2019).

2008   Gazprom Neft, Rosneft, LUKOIL, TNK-BP and Surgutneftegas signed a MoU to begin cooperation and joint participation in projects in Cuba and Venezuela as part of the National Oil Consortium (Gazprom 2019).

2008   Roman Abramovich resigned as Governor of Chukotka (RIA 2008).

2009   Gazprom Neft acquired Naftna Industrija Srbije (NIS) (Gazprom 2019).

2009   Gazprom Neft acquired a controlling share in Sibir Energy as a result of increasing its interest in the Moscow refinery and gaining a stake in the Salym Petroleum Development, a joint venture with Shell (Gazprom 2019).

2009   Gazprom Neft completed the acquisition of the Chevron Italia lubricants plant in Bari (Gazprom 2019).

2010   Gazprom Neft closed a deal to develop the Badra field in Iraq (Gazprom 2019).

2010   Gazprom Neft was appointed to lead the National Oil Consortium in Venezuela (Gazprom 2019).

2010   Gazprom Neft bought a retail network of 20 fuel stations and plots in Kazakhstan and STS-Service, a subsidiary of Sweden's Malka Oil (Gazprom 2019).

2011   The Russian antimonopoly service fined Gazprom Neft USD 29 million as part of the third wave of antitrust cases against Russian oil majors (RAPSI 2011).

2011   Gazprom Neft began the rebranding of its retail network in Europe (Serbia and Romania) under the Gazprom brand (Gazprom 2019).

2011   The 'trial of the century' was held in London: Boris Berezovskiy versus Roman Abramovich. In the trial, detailed information was disclosed about the privatization of Sibneft during the 1990s. Boris Berezovskiy, the plaintiff, lost the trial (BFM.ru 2011).

2012   The authorities of Yamal-Nenets Autonomous District filed six administrative lawsuits with a local court against Gazpromneft-Noyabrskneftegaz for breaking laws at its oil wells (RAPSI 2012a).

2012   Corruption issues related to Gazprom Neft were made public through the leak of the US diplomatic correspondence (The Moscow Times 2012).

2012   The Commercial Court of Appeals confirmed the cancellation of the permit of the Moscow refinery owned by Gazprom Neft to release pollutants into the air (RAPSI 2012b).

2012   Greenpeace activists boarded the Prirazlomnaya oil platform in the Pechora Sea to protest against oil extraction in the Arctic (Greenpeace 2013).

2012   Gazprom Neft Nizhny Novgorod was fined USD 10 000 for acquiring new assets without duly notifying the antimonopoly authority (RAPSI 2012c).

2013   Greenpeace activists boarded the Prirazlomnaya platform for the second time (Greenpeace 2013).

2013   Greenpeace Russia filed a complaint with the Prosecutor General's Office about oil production on Gazprom Neft's Prirazlomnaya oil platform, claiming that it violated several environmental protection and safety laws (RAPSI 2013a).

2013   The Serbian police detained four people at the Gazprom Neft subsidiary Pancevo oil refinery, including one of the directors, on suspicion of corruption (RAPSI 2013b).

2013   The Moscow Commercial Court dismissed a claim filed by BMW contesting the Russian patent regulator's decision to grant legal protection to the G-Drive brand registered by Gazprom Neft (RAPSI 2013c).

2013   The Gazprom Neft board of directors approved the Gazprom Neft 2025 Development Strategy (Gazprom 2019).

2013   Gazprom Neft produced the first ever Russian offshore Arctic oil from the Prirazlomnoye field in the Pechora Sea (Gazprom 2019).

2014   Gazprom Neft's Moscow refinery was fined RUB 100 000 (about USD 1700) for environmental damage (RAPSI 2014a).

2014   EU financial sanctions were introduced against Gazprom Neft (and other Russian oil and defence companies) in connection with the conflict in Ukraine (RAPSI 2014c).

2014   Gazprom Neft filed a claim with the EU Court of Justice against the EU sanctions targeting the company (RAPSI 2014b).

2014   The Greenpeace ship *Arctic Sunrise* left Murmansk after ten months in detention over its protests against the Prirazlomnaya oil platform (RAPSI 2014d).

2014   The Dutch police arrested 30 Greenpeace activists for blocking the Saturn oil rig chartered by Gazprom Neft at the Dutch port of Ijmuiden (RAPSI 2014e).

2014   Production commenced at the Novoportovskoye field with nine wells drilled during 2014. The first shipments of crude were sent to European customers by tanker (Gazprom 2019).

2014   Commercial oil production commenced from the Badra field in Iraq (Gazprom 2019).

2014   Gazprom Neft became the first company in Russia to carry out exploration using supposedly 'green' seismic technology, protecting extensive areas from deforestation (Gazprom 2019).

2015   Gazprom Neft requested RUB 198 billion (about USD 3.3 billion) in financial assistance from the government due to sanctions and the oil price fall (The Moscow Times 2015).

2015   Gazprom Neft was listed as Russia's best employer in the 2015 Russian Employers' Rankings by HeadHunter.ru (Gazprom 2019).

2015   Gazprom Neft acquired the licence to develop the Zapadno-Yubileynoye field in Yamal-Nenets Autonomous District and several licences in Khanty-Mansi Autonomous District (Gazprom 2019).

2016   Gazprom Neft-Aero filed a lawsuit at the Commercial Court of Saint Petersburg for RUB 6 billion (USD 78 million) to be recovered from Transaero airline in its bankruptcy proceedings (RAPSI 2016).

2016   Gazprom Neft failed to pay the salaries of over 150 employees in Yamal (RBC.ru 2016).

2016   The Arctic Gates marine oil terminal was commissioned in the Gulf of Ob, launching the full-scale development of the Novoportovskoye field (Gazprom 2019).

2017   The authorities detained the head of a Gazprom Neft subsidiary for two months in connection with a bribe in the town of Salekhard (RAPSI 2017a).

2017   The Deputy CEO of Gazprom Neft-Hantos was detained in Khanty-Mansi Autonomous District over the embezzlement of RUB 23 million (about USD 400 000) (RAPSI 2017b).

2017   An investigation was conducted into the alleged embezzlement of RUB 2 billion (USD 34 million) from Gazprom Neft (RAPSI 2017c).

2017 Gazprom Neft became the third-largest oil producer in Russia (Gazprom 2019).

2017 Gazprom Neft discovered the offshore Neptune field in the Sea of Okhotsk near Sakhalin (Gazprom 2019).

2017 Gazprom Neft launched its digital transformation programme (Gazprom 2019).

2017 Gazprom Neft commissioned the Biosphere high-tech wastewater treatment facility at its Moscow refinery (Gazprom 2019).

2018 Gazprom Neft raised its oil output by 50 000 barrels per day in 2018 despite the OPEC+ agreement about oil production cuts (Reuters 2018a, 2018b).

2018 Gazprom Neft discovered the Triton field in the Sea of Okhotsk (Neftegaz 2018).

2019 Gazprom Neft was granted the right to carry out exploration and development in three blocks in Western Siberia (Pipeline 2019).

REFERENCES

BFM.ru (2011), 'Kak "Sibneft" prevratilas v "Gazprom Neft"', accessed 10 February 2019 at https://www.bfm.ru/news/160670.

Gazprom (2019), 'History', accessed 10 February 2019 at http://www.gazprom-neft .com/company/history/.

Gazprom Neft (2019), 'Management board', accessed 10 February 2019 at https://www .gazprom-neft.ru/company/management/management-board/#dyukov.

Greenpeace (2013), 'Aktivisty Greenpeace podnyalis na platformu Prirazlomnaya', accessed 10 February 2019 at https://www.greenpeace.org/russia/ru/news/2013/18 -09-action-on-Prirazlomnaya/.

Institute of Energy for South East Europe (2018), 'Gazprom Neft's 1st half 2018 net profit almost doubles', accessed 12 February 2019 at https://www.iene.eu/gazprom -nefts-1st-half-2018-net-profit-almost-doubles-p4483.html.

Kommersant (2001), 'Analitik otvetil za banditov', accessed 10 February 2019 at https://www.kommersant.ru/doc/289985.

Live Journal (2017), 'Noyabrsk. Kak dobyvalas neft i delilsya SSSR', accessed 12 February 2019 at https://avro-live.livejournal.com/168537.html.

Neftegaz (2005), 'Sibneft settled tax claims', accessed 26 April 2020 at https://neftegaz .ru/en/news/companies/421823-sibneft-settled-tax-claims/.

Neftegaz (2018), 'Triton – syn Neptuna!', accessed 10 February 2019 at https:// neftegaz.ru/news/view/176861-Triton-syn-Neptuna-Gazprom-neft-otkryla-2-e -mestorozhdenie-na-shelfe-Ohotskogo-morya.

Pipeline (2019), 'Gazprom Neft wins Western Siberian exploration blocks', accessed 12 February 2019 at https://www.pipelineoilandgasnews.com/regionalinternational -news/international-news/2019/january/gazprom-neft-wins-west-siberian -exploration-blocks/.

RAPSI (2011), 'Antimonopoly watchdog fines Gazprom $29 mln', accessed 10 February 2019 at http://www.rapsinews.com/judicial_news/20111227/259236918 .html.

RAPSI (2012a), 'Gazprom Neft subsidiary suspected of illegal well operation', accessed 10 February 2019 at http://www.rapsinews.com/judicial_news/20121001/ 264824038.html.

RAPSI (2012b), 'Oil refinery's appeal on hazardous emissions dismissed', accessed 10 February 2019 at http://www.rapsinews.com/judicial_news/20120425/262942300 .html.

RAPSI (2012c), 'Gazprom Neft subsidiary fined for purchasing assets', accessed 10 February 2019 at http://www.rapsinews.com/judicial_news/20120418/262847087 .html.

RAPSI (2013a), 'Greenpeace requests inquiry into Arctic oil production', accessed 10 February 2019 at http://www.rapsinews.com/news/20131224/270235764.html.

RAPSI (2013b), 'Serbian police detain head of Gazprom Neft refinery', accessed 10 February 2019 at http://www.rapsinews.com/anticorruption_news/20131114/ 269650470.html.

RAPSI (2013c), 'BMW loses Moscow court action over G-Drive brand', accessed 10 February 2019 at http://www.rapsinews.com/judicial_news/20131011/269170104 .html.

RAPSI (2014a), 'Moscow refinery fined 100,000 roubles for environmental damage', accessed 10 February 2019 at http://www.rapsinews.com/news/20141231/ 272905850.html.

RAPSI (2014b), 'Gazprom Neft files claim with EU Court of Justice over sanctions', accessed 10 February 2019 at http://www.rapsinews.com/judicial_news/20141029/ 272482712.html.

RAPSI (2014c), 'New EU sanctions target Russia's oil and defense companies', accessed 10 February 2019 at http://www.rapsinews.com/news/20140912/272099330.html.

RAPSI (2014d), 'Detained Greenpeace icebreaker to leave Murmansk soon', accessed 10 February 2019 at http://www.rapsinews.com/news/20140721/271763426.html.

RAPSI (2014e), 'Dutch police arrest Greenpeace activists blocking Gazprom oil rig', accessed 10 February 2019 at http://www.rapsinews.com/news/20140527/ 271416547.html.

RAPSI (2016), 'Gazprom Neft subsidiary demands $78 million from Transaero airline', accessed 10 February 2019 at http://www.rapsinews.com/judicial_news/ 20160208/275382673.html.

RAPSI (2017a), 'CEO of Gazprom Neft subsidiary detained for two months in bribery case', accessed 10 February 2019 at http://www.rapsinews.com/judicial_news/ 20171120/280994778.html.

RAPSI (2017b), 'Deputy CEO of Gazprom Neft subsidiary detained on embezzlement charges', accessed 10 February 2019 at http://www.rapsinews.com/judicial_news/ 20170906/280057557.html.

RAPSI (2017c), 'Probe opened into alleged embezzlement at Gazprom Neft – report', accessed 10 February 2019 at http://www.rapsinews.com/judicial_news/20170209/ 277757891.html.

RBC.ru (2016), 'Na Yamale svyshe 150 vakhtovikov neskolko mesyatsev ne poluch- ayut zarplatu', accessed 10 February 2019 at https://t.rbc.ru/tyumen/20/04/2016/ 57173d7f9a79470473771cbb.

Reuters (2018a), 'Russia's Gazprom Neft: Output cuts should be eased this summer', accessed 10 February 2019 at https://af.reuters.com/article/energyOilNews/ idAFR4N1LW03J.

Reuters (2018b), 'Russia's Gazprom Neft sticking with plan to raise oil output in 2019', accessed 12 February 2019 at https://www.reuters.com/article/us-oil-opec-russia -gazpromneft/russias-gazprom-neft-sticking-with-plan-to-raise-oil-output-in-2019 -idUSKBN1O41X1.

RIA (2008), 'Nasledstvo Abramovicha: Chto oligarch ostavil Chukotke', accessed 10 February 2019 at https://ria.ru/20080703/112904809.html.

Sibneft (2019), 'About', accessed 10 February 2019 at http://www.ngfr.ru/library.html ?sibneft.

The Moscow Times (2012), 'Why Gazprom resembles a crime syndicate', accessed 10 February 2019 at https://themoscowtimes.com/articles/why-gazprom-resembles-a-crime-syndicate-12914.

The Moscow Times (2015), 'Russia's Gazprom Neft asks government for financial aid, report says', accessed 10 February 2019 at https://themoscowtimes.com/articles/russias-gazprom-neft-asks-government-for-financial-aid-report-says-44532.

Vedomosti (2011), 'Sibneft byla kuplena po korruptsionnoy skheme', accessed 10 February 2019 at https://www.vedomosti.ru/business/articles/2011/10/06/priznalsya _v_korrupcii.

Appendix 4: Corporate history of Surgutneftegas

Early 20th century	The history of petroleum product distributors Novgorodnefteprodukt and Tvernefteprodukt that are now part of Surgutneftegas dates back to the early twentieth century and is connected with the Nobel brothers, who established the first Russian petroleum trading enterprises (NGFR 2006; Surgutneftegas 2017).
1960s	The territory along the Ob River near the city of Surgut became one of the first areas of oil and gas production in Western Siberia (Surgutinfo 2017).
1961	The construction of the Kirishi refinery began. In December 1965, the refinery received the first oil delivery, and four months later, the first petroleum products were delivered (NGFR 2006).
1969	The Yaroslavl-Kirishi pipeline was commissioned, and Kirishinefteorgsintez began to receive oil from the West Siberian fields (NGFR 2006).
1984	Vladimir Bogdanov, who was 33 at the time, was appointed General Director of Surgutneftegas Production Association (Forbes 2018).
1992	Presidential Decree 1403 established Surgutneftegas (along with LUKOIL, YUKOS, Rosneft, Transneft and Transnefteproduct) (Kremlin 1992).
1993	Surgutneftegas was registered by the administration of the town of Surgut (NGFR 2006).
1993	The corporatization and privatization of Surgutneftegas began: 8% of its shares were sold at a closed auction, and 7% were bought by the company for vouchers (NGFR 2006).
1993	The open joint stock company (OJSC) Surgutneftegas was established by Resolution 271 of the Council of Ministers of the Russian Federation. The following previously state-owned enterprises were included in the company: Surgutneftegas, Kirishinefteorgsintez, Ruchyi and Krasnyi neftyanik oil storage depots, Kirishskoye petroleum products supply company, Karelnefteprodukt, Novgorodnefteprodukt, Pskovnefteprodukt, Tvernefteprodukt, Kaliningradnefteprodukt, Petersburgneftesnab and Saint Petersburg Industrial Automotive Service Plant (NGFR 2006).
1996	Surgutneftegas developed a programme for the reorganization of its subsidiaries and share swaps. However, it had to be postponed due to stiff opposition from the minority shareholders of these companies (NGFR 2006).
1996	The Russian government issued a resolution granting Surgutneftegas the rights to build and operate the oil terminal in the Batareinaya Bay. It also obtained a special permit to export an additional 5 million tonnes of oil to finance the project (the port was not eventually built) (Dp.ru 2017; NGFR 2006).

1997	Surgutneftegas acquired 14.99% of Nafta-Moskva, a major oil exporter. Later the share was increased to a 15% blocking stake (NGFR 2006).
1997	Surgutneftegas issued level-1 American depository receipts (ADRs) (NGFR 2006).
1997	The Gunvor oil trader was established. It was closely connected with Gennadiy Timchenko and sold large volumes of Surgutneftegas crude and petroleum products that were previously sold through Kineks, also controlled by Timchenko (Rupres 2008).
1998	Surgutneftegas sold its subsidiaries in Saint Petersburg because of conflicts with the local authorities (Kommersant 1998).
1998	Surgutneftegas was the only Russian oil company to increase its oil production during the financial crisis (NGFR 2006).
1999	Surgutneftegas reported the largest increase in oil production in the industry. During 1999, the company's shares appreciated by 460% (Newsrus 2010).
2000	Surgutneftegas demonstrated the highest profitability among Russian oil companies and the largest rise in cash flows while its operating expenses remained the lowest (Neftegas.ru 2002).
2000	Vladimir Bogdanov acted as the authorized representative of Vladimir Putin during his first presidential election (Kommersant 2012c).
2000	Transneft accused Surgutneftegas of dubious oil exports taking advantage of the concessional terms for Russian petroleum products (Mokrousova 2012).
2001	Surgutneftegas purchased the Surgut gas-processing plant from SIBUR-Tyumen, and the plant was integrated into its gas-processing division (Golubeva and Rodina 2017).
2001	Surgutneftegas started installing gas-powered turbines in Tyanskoye and Konitlorskoye fields to supply them with electricity using APG (Uralinformbyuro 2001).
2001	An office was created within Surgutneftegas to introduce new technologies (side-tracking of wells with high water-cut and low-yield wells) to reduce the oil production decline in the oldfields (NGFR 2006).
2001	The stake in Nafta-Moskva was sold to Suleiman Kerimov, and Surgutneftegas began exporting via its own subsidiary company (Butrin 2001).
2001	Surgutneftegas issued its first financial report in accordance with US GAAP standards. This disclosed the amount of its treasury shares as 46.6% of its voting stock (Fundinguniverse 2017; Neftgas.ru 2017).
2001	The largest minority shareholders of Surgutneftegas sued the company, demanding that Surgut not vote with its treasury stock (Fundinguniverse 2017; Neftgas.ru 2017).
2001	Minority shareholders of Surgutneftegas filed a lawsuit against the company demanding higher dividends (Fundinguniverse 2017; Neftgas.ru 2002).
2003	Surgutneftegas acquired the rights to the major Talakanskoye field in the Republic of Sakha (Yakutiya), which had been held by YUKOS (News.ru 2003).
2003	Surgutneftegas acquired licences for geological exploration at the Khorokorskoye, Verkhnepolidinskoye and Kedrovoye areas adjacent to the Talakanskoye field (NGFR 2006).
2003	Gazprom, Rosneft and Surgutneftegas signed an agreement defining Eastern Siberia and the Republic of Sakha as an area of shared interest (Pravda.ru 2003; Zatologin and Ustinov 2004).

2003	Surgutneftegas Oil Company was transformed into LLC Leasing Production to block attempts to gain control over Surgutneftegas by external investors (Kuleshov 2003; NGFR 2006).
2004	During the auction for Yuganskneftegaz, two managers of Surgutneftegas represented Baikal Finance Group, the winner of the auction (Vedomosti 2014).
2004	Surgutneftegas acquired a 100% stake in the authorized capital of Kondaneft (NGFR 2006).
2004	Surgutneftegas paid eight times more taxes per tonne of oil than Sibneft and three times more than TNK (Ignatova 2003).
2004	The Arbitration Court of Khanty-Mansi Autonomous District rejected a claim by minority shareholders against Surgutneftegas, seeking redemption of 62% of the shares controlled by the company (Gazeta.ru 2013; NGFR 2006).
2005	Surgutneftegas began working on the Bazhenov formation.
2005	Hermitage Capital Management, once Russia's largest foreign portfolio investor, filed a legal complaint to force Surgut to disclose its ownership structure. Shortly afterwards, William Browder, its head, was banned from entering Russia 'for reasons of national security' (Latinina 2008).
2005	The tax accounting of Surgutneftegas was transferred to Moscow (Finam.ru 2017a).
2005	Severstal sold Surgutneftegas 35% of Ren TV (Vedomosti 2005).
2006	Surgutneftegas sold LLC Production Leasing to an unknown buyer (Vedomosti 2007).
2006	Surgutneftegas's workers began to demand an increase in the guaranteed part of their wages. The company announced a 20% raise, but it was in fact just a small increase by a few hundred roubles. Trade unions organized several large protests (Kommersant 2006).
2006	Surgutneftegas oil production began to decline, and this continued for five years.
2007	The newspaper *Vedomosti* uncovered the complex ownership structure of Surgutneftegas (Derbilova and Mazneva 2007).
2007	The Russian political analyst Stanislav Belkovsky claimed that Vladimir Putin effectively controls 37% of the shares of Surgutneftegas (Guardian 2007).
2007	Several Russian news agencies received a fake message about the arrest of General Director Vladimir Bogdanov and chairman of the board of directors Nikolai Zakharchenko on suspicion of tax evasion, signed by the Surgutneftegas press centre (Mazneva 2007a).
2008	Surgutneftegas commissioned Talakanskoye field in Yakutiya (Neftegazovaya vertikal 2014).
2008	Surgutneftegas joined the National Oil Consortium in Venezuela with a 20% stake (Kommersant 2012a).
2009	Surgutneftegas raised its stake in National Media Group from 12.3% to 24% (Kommersant 2009).
2009	Surgutneftegas bought 21.2% of Hungary's MOL from the Austrian OMV (Vedomosti 2009).
2009	Surgutneftegas commissioned the Alinskoye field in the Republic of Sakha (Neftegazovaya vertikal 2014).
2011	Surgutneftegas accumulated almost USD 20 billion in its year-end savings accounts (Mazneva 2007b).
2011	Surgutneftegas sold its stake in MOL to the government of Hungary (Kommersant 2011).

2011	Surgutneftegas oil production began to rise, mainly due to the Talakanskoye and Alinskoye fields in Yakutiya (Burneft 2017).
2011	Surgutneftegas demonstrated the best drilling results among the Russian oil companies (Mazneva 2007b).
2012	Surgutneftegas withdrew from the Russian National Oil Consortium developing Venezuela's Junin-6 project. Its stake was bought by Rosneft in 2013 (Reuters 2012).
2012	Surgutneftegas reported savings of USD 30 billion (Russianrt.com 2014).
2012	Surgutneftegas bought a licence for the Shpielman oilfield in Khanty-Mansi Autonomous District (Kommersant 2012b).
2013	For the first time since 2001, Surgutneftegas resumed publishing its accounts in accordance with international financial reporting standards (IFRS), complying with a 2011 law applicable to public companies (Seeking Alpha 2016).
2014	Surgutneftegas commissioned three new fields in Western Siberia – Vysotnoye, Kochevskoye and Verkhnekazymskoye (Glavportal 2017).
2014	The United States introduced sanctions against Russian companies, including Surgutneftegas, in connection with the conflict in Ukraine (Forbes 2014).
2014	Mikhail Khodorkovskiy accused Surgutneftegas of being affiliated with the Russian authorities (Osipov 2014; Stepigin 2014).
2015	Surgutneftegas rose six places to number 12 in the world in the S&P Platts Top 250 Global Energy Company Rankings (SPglobal 2015).
2015	According to Bloomberg, Surgutneftegas was the only one of the largest oil companies in the world that continued to generate income for investors after the collapse of oil prices in 2014 (Forbes 2017; Kalukov and Pobedova 2016).
2015	Surgutneftegas reported one of the lowest salary levels among Russian oil companies – an average of RUB 51 772 (Utmag 2015).
2015	Surgutneftegas commissioned the Shpielman oilfield (Vedomisti 2015).
2015	In Khanty-Mansi Autonomous District, Surgutneftegas sued the local shaman of the Khanty ethnic group, accusing him of death threats (Tumanov 2015).
2016	Surgutneftegas denied being interested in buying the shares of Bashneft and Rosneft (Derbisheva 2016; Finam.ru 2016a).
2016	Vladimir Putin awarded the title Labour Hero of Russia to the Head of Surgutneftegas, Vladimir Bogdanov (Finam.ru 2017b).
2016	The board of directors of Surgutneftegas approved a new dividend policy (Finam 2016b).
2016	Vassilyi Pyak, a resident of Khanty-Mansi Autonomous District, reported that Surgutneftegas was going to start oil production within the Numto Nature Park. After a campaign by Greenpeace, public hearings were held, and the re-zoning plan was amended; however, protests still continue against oil drilling in the area (Fedpress.ru 2015).
2017	Surgutneftegas accumulated USD 34 billion in its accounts (Neftyanka 2017).
2017	Rumours appeared about the possible changes in Surgutneftegas, including the retirement of Vladimir Bogdanov and the relocation of the company's headquarters to Moscow (Neftyanka 2017).

2017	Surgutneftegas announced a decision to insure the company's top management for up to USD 50 million (Neftyanka 2017).
2017	Surgutneftegas for the first time since 1995 reported a net financial loss for the year (Vedomosti 2017).
2017	Surgutneftegas was ranked only 165th in the S&P Global Platts Top 250 Global Energy Company Rankings (a dramatic fall from number nine in 2016) (Global Energy Company Rankings 2018).
2017	Surgutneftegas postponed the launch of new projects in compliance with the agreement with OPEC.
2018	Surgutneftegas participated in the agreement between the Russian government and Russian oil companies to freeze oil prices until 31 December 2018 (RBC 2018).
2019	Dividend payments by Surgutneftegas fell by 10–15% compared to the previous year (Finam. ru 2019).

REFERENCES

Burneft.ru (2017), 'OAO Surgutneftegaz: Ne po shablonu', accessed 12 March 2019 at https://burneft.ru/archive/issues/2017-09/24.

Butrin, D. (2001), 'Nafta-Moskva raschitaet aksionerov', accessed 3 October 2017 at https://www.kommersant.ru/doc/294065.

Derbilova, E. and E. Mazneva (2007), 'Tayniki Surguta', accessed 3 October 2017 at https://www.vedomosti.ru/newspaper/articles/2007/01/24/tajniki-surguta.

Derbisheva, N. (2016), 'Surgutneftegaz zayavil ob otsutstvii interesa k Bashnefti i Rosnefti', accessed 3 October 2017 at https://www.rbc.ru/business/29/06/2016/577380019a7947901d3c8883.

Dp.ru (2017), 'Portovyye skandaly i operatsiya chistyye ruki: O chem pisal DP 20 let nazad', accessed 21 September 2017 at https://www.dp.ru/a/2017/06/23/Portovie_skandali_i_opera.

Fedpress.ru (2015), 'Tam vodyatsya sterkhi: Putina prosyat pomoch sokhranit Numto', accessed 3 October 2017 at http://fedpress.ru/article/1747054.

Finam.ru. (2016a), 'Otkaz Surgutneftegaza ot prityazaniy na Bashneft sposoben negativno skazatsya na tsene aktsiy obeikh kompaniy', accessed 3 October 2017 at https://www.finam.ru/analysis/marketnews/vcherashniiy-otskok-po-fondovym-indeksam-i-syryu-pozvolyaet-rasschityvat-na-to-chto-nekontroliruemogo-obvala-izbezhat-udalos-20160629-12000/.

Finam.ru. (2016b), 'Sovet direktorov Surgutneftegaza utverdil dividendnuyu poli-tiku', accessed 3 October 2017 at https://www.finam.ru/analysis/newsitem/sovet-direktorov-surgutneftegaza-utverdil-dividendnuyu-politiku-20161003-114726/.

Finam.ru. (2017a), 'Surgutneftegas', accessed 3 October 2017 at https://www.finam.ru/analysis/newsitem18AED/.

Finam.ru. (2017b), 'Putin prisvoil Surgutneftegasa zvanie geroya truda', accessed 3 October 2017 at https://www.finam.ru/analysis/newsitem/putin-prisvoil-glave-surgutneftegaza-zvanie-geroya-truda-20160421-153348/.

Finam.ru (2019), 'Dividendy po aktsiyam "Surgutneftegaza" po itogam 2019 goda budut na 10–15% nizhe, chem za 2018 god', accessed 11 February 2019 at https://

www.finam.ru/analysis/marketnews/dividendy-po-akciyam-surgutneftegaza-po
-itogam-2019-goda-budut-na-10-15-nizhe-chem-za-2018-god-20190207-18000/.

Forbes (2014), 'SSHA vveli sanktsii protiv Lukoyla, Gazproma i Sberbanka', accessed
12 March 2019 at https://www.forbes.ru/news/267601-ssha-vveli-novye-sanktsii
-protiv-rossii.

Forbes (2017), 'Surgutneftegas', accessed 3 October 2017 at http://www.forbes.ru/
profile/244786-surgutneftegaz?from_rating=327357.

Forbes (2018), 'Vladimir Bogdanov', accessed 12 March 2019 at https://www.forbes
.ru/profile/vladimir-bogdanov.

Fundinguniverse (2017), 'OAO Surgutneftegaz history', accessed 3 October 2017 at
http://www.fundinguniverse.com/company-histories/oao-surgutneftegaz-history/.

Gazeta.ru (2013), 'Surgut nichem ne riskuet', accessed 23 September 2017 at https://
www.gazeta.ru/business/2013/04/30/5286761.shtml.

Glavportal (2017), 'Surgutneftegas: 40 let pobed', accessed 3 October 2017 at https://
glavportal.com/materials/surgutneftegaz-40-let-pobed/.

Global Energy Company Rankings (2018), '2018 top 250 companies', accessed 11
February 2019 at https://top250.platts.com/Top250Rankings.

Golubeva, I and E. Rodina (2017), 'Gazopererabatyvayushchiye predpriyat-
iya rossii', accessed 8 October 2019 at http://npirf.ru/spravochnik-monografiya
-gazopererabatyvayushhie-predpriyatiya-rossii/.

Guardian (2007), 'Putin, the Kremlin power struggle and the $40bn fortune', accessed
3 October 2017 at https://www.theguardian.com/world/2007/dec/21/russia
.topstories3.

Ignatova, M. (2003), 'Surgutskiy pasians', accessed 3 October 2017 at http://www
.forbes.ru/Forbes/issue/2004-04/2365-surgutskii-pasyans.

Kalukov, E. and L. Pobedova (2016), 'Surgutneftegaz stal samoy dokhodnoy dlya
investorov neftyanoy kompaniyey', accessed 3 October 2017 at https://www.rbc.ru/
business/19/02/2016/56c6d3ac9a7947bdc207ece2.

Kommersant (1998), 'Surgutneftegaz ukhodit iz Peterburga', accessed 12 March 2019
at https://www.kommersant.ru/doc/202910.

Kommersant (2006), 'Surgutneftegas prizvali k socialnoy otvetsvennosti', accessed 3
October 2017 at https://www.kommersant.ru/doc/691383.

Kommersant (2009), 'Natsionalniye media napolnilis Surgutneftegasom', accessed 12
March 2019 at https://www.kommersant.ru/doc/1129127.

Kommersant (2011), 'Surgutneftegaz otstupil bez poter', accessed 12 March 2019 at
https://www.kommersant.ru/doc/1647101.

Kommersant (2012a), 'Vladimir Bogdanov ne srabotalsya s venesuelskoy neftyu',
accessed 12 March 2019 at https://www.kommersant.ru/doc/2030559.

Kommersant (2012b), 'Surgutneftegaz nachinayet tratit', accessed 12 March 2019 at
https://www.kommersant.ru/doc/2093757.

Kommersant (2012c), 'Doveriye, ot kotorogo trudno otkazatsya', accessed 12 March
2019 at https://www.kommersant.ru/doc/1867358.

Kremlin (1992), 'Ukaz Prezidenta Rossiyskoy Federatsii', accessed 12 March 2019 at
http://www.kremlin.ru/acts/bank/2417/page/1.

Kuleshov, I. (2003), 'Surgutneftegas likvidirovali', accessed 3 October 2017 at https://
www.kommersant.ru/doc/424475.

Latinina, J. (2008), 'Martishki s administrativnim resursom', accessed 3 October
2017 at https://www.novayagazeta.ru/articles/2008/11/27/35702-martyshki-s
-administrativnym-resursom.

Mazneva, E. (2007a), 'Hakery obvinili Bogdanova v neuplate nalogov', accessed 3 October 2017 at https://www.vedomosti.ru/library/articles/2007/07/04/hakery -obvinili-bogdanova-v-neuplate-nalogov.

Mazneva, E. (2007b), 'Stragegia Surgutneftegasa ostaetsya prejnej – burit, burit, burit', accessed 3 October 2017 at https://www.vedomosti.ru/business/articles/2012/06/29/ strategiya-surgutneftegaza-ostaetsya-prezhnej-burit-burit-burit.

Mokrousova, I. (2012), *Druzya Putina*, Moskva: EKSMO.

Neftegas.ru (2002), 'Surgutneftegas: Perviy otchet', accessed 3 October 2017 at https:// neftegaz.ru/analisis/view/5804-Surgutneftegaz-pervyj-otchet.

Neftegazovaya vertikal (2014), 'Surgutneftegaz v yakutii: Desyat let spustya', accessed 12 March 2019 at http://www.ngv.ru/upload/iblock/07c/07caf7195 802fc8439112aaebd08ce09.pdf.

Neftyanka (2017), 'Strategii kompaniy. Surgutneftegaz: Riski rastut', accessed 11 February 2019 at http://neftianka.ru/strategii-kompanij-surgutneftegaz-riski-rastut/.

News.ru (2003), 'Mestorozhdeniye YUKOSa otdali Surgutneftegazu', accessed 12 March 2019 at https://www.newsru.com/finance/03nov2003/yuk.html.

Newsrus (2010), 'Istoriya Surgutneftegaza', accessed 8 October 2019 at http:// newsruss.ru/doc/index.php/.

NGFR (2006), 'Surgutneftegas', accessed 11 September 2017 at http://www.ngfr.ru/ library.html?rosneft.

Osipov, I (2014), 'Mikhail Khodorkovskiy: "Okruzheniyu Putina ya ne doveryayu i na pyat kopeyek"', accessed 3 October 2017 at http://www.forbes.ru/sobytiya/ obshchestvo/263999-mikhail-khodorkovskii-okruzheniyu-putina-ya-ne-doveryayu -i-na-pyat-kopee.

Pravda.ru (2003), 'Gazprom, Rosneft i Surgutneftegaz osvoyat Vostochnuyu Sibir bez YUKOSa', accessed 3 October 2017 at https://www.pravda.ru/news/economics/ 12219-gazprom_rosneft_surgutneftegaz_osvoenie_mestorozhdenii/.

RBC (2018), 'Surgutneftegaz prisoyedinilsya k soglasheniyu o zamorozke tsen na toplivo', accessed 11 February 2019 at https://www.rbc.ru/rbcfreenews/ 5bdaa5539a794743fcc0f0c0.

Reuters (2012), 'Russia's Surgut says leaves Venezuela oil consortium', accessed 3 October 2017 at https://uk.reuters.com/article/surgut-venezuela/russias-surgut-says -leaves-venezuela-oil-consortium-idUKL5E8M734Q20121107.

Rupres (2008), 'Ataka na Gennadiya Timchenko', accessed 12 March 2019 at https:// www.rospres.org/finance/2072/.

Russianrt.com (2014), 'Die Welt: Surgutneftegas – taejnaya tajna Putina', accessed 3 October 2017 at https://russian.rt.com/inotv/2014-07-08/Die-Welt-Surgutneftegaz ---taezhnaya.

Seeking Alpha (2016), 'Surgutneftegaz: Even in Siberia there is happiness', accessed 3 October 2017 at https://seekingalpha.com/article/4030315-surgutneftegaz-even -siberia-happiness.

SPglobal (2015), 'Media Center', accessed 3 October 2017 at https://www.spglobal .com/platts/en/about-platts/media-center/press-releases.

Stepigin, A. (2014), 'Zagovoril. Mikhail Khodorkovskiy peredal privet Vladimiru Bogdanovu cherez Forbes ...', accessed 3 October 2017 at https://ura.news/news/ 1052186396.

Surgutinfo (2017), 'OAO Surgutneftegas', accessed 18 September 2017 at http://www .v-surgut.ru/surgutneftegaz.shtml.

Surgutneftegas (2017), 'Istoriya kompanii', accessed 18 September 2017 at http:// surgutneftegas.ru/en/about/history/.

Tumanov, G. (2015), 'Shamannoye pravo', accessed 3 October 2017 at https://www .kommersant.ru/doc/2791298.

Uralinformbyuro (2001), 'Surgutneftegaz obespechivayet sebya elektroenergiyey', accessed 12 March 2019 at https://www.uralinform.ru/news/economy/2956 -surgutnephtegaz-obespechivaet-sebya-elektroenergiei/.

Utmag (2015), 'Finansoviye bitvi: Surgutneftegas vs Magnit', accessed 3 October 2017 at https://utmagazine.ru/posts/12466-finansovye-bitvy-surgutneftegaz-vs-magnit.

Vedomosti (2005), 'Surgut popal na Ren TV', accessed 12 March 2019 at https://www .vedomosti.ru/newspaper/articles/2005/09/02/surgut-popal-na-ren-tv.

Vedomosti (2007), 'Novaya zagadka Surgutneftegaza', accessed 12 March 2019 at https://www.vedomosti.ru/newspaper/articles/2007/01/10/novaya-zagadka -surgutneftegaza.

Vedomosti (2009), 'Ni vrag, ni drug', accessed 12 March 2019 at https://www .vedomosti.ru/newspaper/articles/2009/04/24/ni-vrag-ni-drug.

Vedomosti (2014), 'Za kompaniei, kupivshei Yuganksneftegas v 2004 godu, stoyal Surgutneftegas', accessed 12 March 2019 at https://www.vedomosti.ru/business/ articles/2014/07/28/gaagskij-sud-za-kompaniej-bajkalfinansgrup-kupivshej.

Vedomosti (2015), 'Surgutneftegas vvel v ekspluatatsiyu mestorozhdeniye im. Shpil'mana', accessed 3 October 2017 at https://www.vedomosti.ru/business/news/ 2015/09/03/607392-surgutneftegaz-vvel-shpilmana.

Vedomosti (2017), 'Surgutneftegaz vpervyye s 1995 g. poluchil chistyy ubytok za god', accessed 11 February 2019 at https://www.vedomosti.ru/business/articles/2017/04/ 03/683758-surgutneftegaz-pervii-ubitok.

Zatologin, D. and E. Ustinov (2004), 'Kupit nelzya prodat', accessed 3 October 2017 at http://www.old.rcb.ru/Archive/articles.asp?id=4617.

Appendix 5: Corporate history of Tatneft

1948   The super-giant Romashkinskoye field was discovered in Tatarstan (Tatneft 2018a).

1950   Tatneft Group was established by a resolution of the Council of Ministers of the Soviet
        Union, incorporating the Bavlyneft oil company, the Bugulmaneft oil-producing trust, the
        Tatburneft drilling company, the Tatneftepromstroy construction and installation trust and the
        Tatnefteproyekt design office (Tatneft 2018a).

1951   The Drilling Personnel School, which had previously been part of the Saratovneftegaz association,
        was transferred to Tatneft (Tatneft 2018a).

1956   Tatneft produced 18 million tonnes of oil and was the largest oil producer in the Soviet Union
        (Tatneft 2018a).

1962   Water injection was used for the first time in the Soviet Union in the Romashkinskoye field.
        A group of Tatneft scientists and officials was awarded the prestigious Lenin Prize for this
        achievement (Tatneft 2018a).

1962   The Tatneftegaz trust and Yelkhovneft Oil Production Board were established (Tatneft 2018a).

1975   The highest ever level of annual oil production in Tatarstan, 104 million tonnes, was reached, and
        gradual decline began afterwards (Tatneft 2018a).

1990   Tatneft's first large-scale environmental programme was launched (Tatneft 2018b).

1994   Agreement on the division of competencies and authority between the Russian Federation and the
        Republic of Tatarstan was signed (Neft i kapital 2007).

1994   Tatneft was corporatized, and its privatization began (Tatneft 2018b).

1995   Tatneft stabilized its oil production (Tatneft 2018b).

1995   The Kichuy refinery with a capacity of 400 000 tonnes of oil per annum was built (Tatneft 2018b).

1996   Tatneft's American depository receipts (ADRs) were listed on the London Stock Exchange
        (Tatneft 2018b).

1997   Tatneft entered the debt market by issuing Eurobonds worth USD 300 million (Vedomosti 2010).

1998   Tatneft's ADRs were listed on the New York Stock Exchange (Tatneft 2018b).

1998   Tatneft's board of directors declared the construction of Nizhnekamsk refinery a strategic priority
        (Tatneft 2018b).

1998   Tatneft bought major blocks of shares in the petrochemical enterprises of Tatarstan (Tatneft
        2018b).

1998   The Kalmtatneft Joint Venture was founded in Kalmykia for exploration and development of
        oilfields (Tatneft 2018b).

2000   The Tatneft Youth Organization was established (Tatneft 2018b).

2002   The management structure of Tatneft was optimized with the establishment of gas (TatNefteGazpererabotka) and petrochemical (TatNeft-NefteKhim) subsidiaries (Tatneft 2018b).

2002   A corporate conflict concerning control over the Moscow refinery erupted between Sibneft and Tatneft, on the one hand, and the Moscow Oil and Gas Company, on the other (Tatneft 2018b).

2002   Tatneft participated in the financing and construction of a bridge across the Kama River (Tatneft 2018b).

2003   The international rating agency Fitch upgraded the rating of Tatneft debt from B- to B. Similarly, S&P upgraded the company's credit rating (Tatneft 2018b).

2003   The thirtieth anniversary of Nizhnekamskshina, the largest enterprise of Tatneft-Neftekhim (Tatneft 2018b).

2003   Tatneft began to develop Tatarstan's resources of extra-viscous oil (Tatneft 2018b).

2004   Tatneft, Nizhnekamskneftekhim, Svyazinvestneftekhim and South Korean LG established the Tatar-Korean Petrochemical Company (TKNK) (Tatneft 2018b).

2004   Tatneft established the Gifted Children Foundation (Tatneft 2018b).

2004   S&P downgraded its rating of Tatneft from B to B- (Vedomosti 2006a).

2005   Tatneft was rated among the top 12 companies in Russia by the *Big Business* magazine. Tatneft was also rated Russia's best company in the category innovation, rationalization and patent/ licensing (Tatneft 2018b).

2005   The Tatarstan Security Council decided to build an oil-refining complex for Tatneft (Tatneft 2018b).

2006   S&P revoked its B- rating of Tatneft (Vedomosti 2006a).

2006   Tatneft began publishing management discussion and analysis (MD&A), thus improving its disclosure ('Tatneft: The conversion of the Tatars' 2014).

2006   Tatneft was the first Russian oil company to downgrade its proved reserves (Vedomosti 2006b).

2006   Tatneft began cooperation with the Russian automaker AvtoVAZ (Tatneft 2018b).

2006   Using a conventional drilling rig, Tatneft drilled a unique U-shaped horizontal through well exiting to the surface at Ashalchinskoye bituminous field (Tatneft 2018b).

2006   Tatneft delisted its ADRs from the New York Stock Exchange (Vedomosti 2006c).

2006   Tatneft obtained more licences than any other company in Libya's third post-sanctions international licensing round. However, when hostilities reignited, it had to leave the country (Soldatkin 2011).

2006   Tatarstan transferred to Tatneft's trust management its stake in Ukrtatnafta, which controlled the Kremenchug refinery in Ukraine (Vedomosti 2006d).

2006   Tatneft's earnings from the sale of technology exceeded RUB 1 billion (Tatneft 2018b).

2007   A corporate raid supported by Ukrainian special forces replaced the leadership of Ukrtatnafta, which had been loyal to Tatneft, and reinstated the former chairman of its managing board (Vedomosti 2007).

2008   The Tatneft-Sibneft alliance won a battle for a controlling interest in the Moscow refinery (Tatneft 2018b).

2008   Tatneft was included in the Platts Top 250 global rating of the most effective energy companies (Tatneft 2018b).

2008	Tatneft and Turkmenneft signed a protocol on cooperation in the petroleum sector in Turkmenistan (Tatneft 2018b).
2008	The cornerstone was laid for a glass fibre-producing plant in the Yelabuga free economic zone (Tatneft 2018b).
2008	The conflict over the Moscow refinery was resolved (Vedomosti 2008a).
2008	Tatneft filed a lawsuit in Zurich against Ukraine demanding reimbursement of USD 1.1 billion for losses incurred because it was forced out of Ukrtatnafta (Vedomosti 2008b).
2009	Tatneft for the third consecutive year came top in the ranking '100 Best Companies in Russia: Ecology and Environmental Management' (Tatneft 2018b).
2010	Tatneft started commercial oil production in Syria (Tatneft 2018b).
2011	Tatneft opened a branch office in Turkmenistan to provide enhanced oil recovery services for the Goturdepe field (Tatneft 2018b).
2011	The management of Tatneft was suspected of corruption and asset stripping (Tarasov 2011).
2011	The first 100 tonnes of extra-viscous oil were produced from Ashalchinskoye field (Tatneft 2018b).
2011	The first stage of TANECO was commissioned (Tatneft 2018b).
2012	Tatneft and the automaker AvtoVAZ signed a strategic partnership agreement supplying Nizhnekamskshina's tyres to Togliatti car plant in 2012–14 (Tatneft 2018b).
2012	Tatneft adopted a new health, safety and environment (HSE) policy (Tatneft 2018b).
2012	Tatneft presented its e-learning projects in Silicon Valley (Tatneft 2018b).
2012	The first batch of TANECO's synthetic oil plant products for AvtoVAZ was delivered (Tatneft 2018b).
2012	Tatneft was recertified, confirming its compliance with ISO 14001:2004 and OHSAS 18001:2007 (Tatneft 2018b).
2014	The shares of Tatneft were included on the quotation list of the highest level of listing at the Moscow Stock Exchange (Tatneft 2018b).
2014	Tatneft registered a German patent for the invention 'Hydrogen sulphide removal unit for gas purification' (Tatneft 2018b).
2015	Tatneft launched a corporate social network (CSN) (Tatneft 2018b).
2015	Tatneft launched a campaign for the collection of used household batteries (Tatneft 2018b).
2016	The Tatneft 2025 Development Strategy was launched (Tatneft 2018b).
2016	Moody's upgraded Tatneft's credit rating to Baa3 with a 'positive' forecast (Taftneft 2018c).
2016	Tatneft was included among the top ten companies of the global petroleum industry by Boston Consulting Group (Vedomosti 2016).
2016	The share of Tatneft in the capital of Zenit Bank exceeded 50% (Tatneft 2018b).
2016	Tatneft was ranked first by Thomson Reuters in a list of Europe's most innovative oil and gas companies (Tatneft 2016).
2016	Tatneft was accused of causing a crisis in Tatfondbank by withdrawing large funds and causing the bank liquidity problems (Inkazan.ru 2016).

2016	Tatneft opened trading on the London Stock Exchange and met with international investors as part of the twentieth anniversary of the company's listing on the London Stock Exchange (Tatneft 2018b).
2017	Tatneft signed an agreement on cooperation with the Ministry of Natural Resources and the Environment of the Russian Federation, the Federal Service for Supervision of Natural Resource Management and the government of the Republic of Tatarstan (Tatneft 2018b).
2017	At Tatneft headquarters in Almetyevsk, seven people were arrested and accused of being members of an international terrorist group (Hizbut-Tahrir). One of them was an employee of Tatneft (Inkazan.ru 2016).
2017	The newspaper *Kommersant* dubbed the legal conflict between Tatneft and Ukraine one of the most important corporate conflicts of 2017 (Kommersant 2017).
2017	Fitch Ratings confirmed Tatneft's credit rating at BBB- with a 'stable' forecast (Tatneft 2017).
2018	Tatneft decided to consolidate its charitable activities into a single unit – the Tatneft Charitable Fund (Tatneft 2018b).
2018	Tatneft tyre-manufacturing companies united under the new corporate brand 'Kama Tyres' (Tatneft 2018b).
2018	RAEX Rating Agency confirmed Tatneft's credit rating at AAA level (Tatneft 2018b).
2018	Moody's upgraded the credit rating of Tatneft to Baa3 with a 'positive' forecast (Tatneft 2018c).
2018	Tatneft was accused of pollution and was obliged to pay around RUB 0.5 million (Tatar-inform 2018).

REFERENCES

Inkazan.ru (2016), 'Krizis v Tatfondbanke sprovotsirovala Tatneft – istochnik', accessed 8 February 2019 at https://inkazan.ru/news/society/19-12-2016/krizis-v-tatfondbanke-sprovotsirovala-tatneft-istochnik.

Kommersant (2017), 'Samiye krupniye sudebniye spory 2017 goda', accessed 12 March 2019 at https://www.kommersant.ru/doc/3427410.

Neft i kapital (2007), 'Bolshaya istoriya', *Neft i kapital*, **8**, 26–31.

Soldatkin, V. (2011), 'Russia Tatneft in $100 mln Libya capex loss-source', accessed 18 September 2017 at http://www.reuters.com/article/russia-tatneft/russia-tatneft-in-100-mln-libya-capex-loss-source-idUSLDE72R0J120110328.

Tarasov, S. (2011), 'Top-menedzhment Tatnefti zapodozrili v vivode aktivov', accessed 8 February 2019 at http://wek.ru/top-menedzhment-tatnefti-zapodozrili-v-vyvode-aktivov.

Tatar-inform (2018), 'PAO Tatneft zaplatilo bolee polumilliona rubley za zagryaznenie okruzhayushey sredy', accessed 8 February 2019 at https://www.tatar-inform.ru/news/2018/08/29/624413/.

'Tatneft: The conversion of the Tatars' (2014), in editor unknown, *Russian Oil and Gas: Two Weddings and a Funeral*, Sberbank Investment Research, February, pp. 35–49.

Tatneft (2016), 'Tatneft vozglavila reyting glavnykh innovatorov mira', accessed 12 March 2019 at http://www.tatneft.ru/press-tsentr/press-relizi/more/4634/?lang=ru.

Tatneft (2017), 'Fitch ratings podtverdilo kreditnyy reyting Tatnefti na urovne BBB-so stabilnym prognozom', accessed 12 March 2019 at http://www.tatneft.ru/press-tsentr/press-relizi/more/5507/?lang=ru.
Tatneft (2018a), 'Starting with the first oilfield to the establishment of Tatneft Company (1943–1990)', accessed 8 February 2019 at http://www.tatneft.ru/about-tatneft/history-of-tatneft-group/starting-with-the-first-oil-field-to-the-establishment-of-tatneft-company-1943---1990?lang=en.
Tatneft (2018b), 'Recent history (1990–2018)', accessed 8 February 2019 at http://www.tatneft.ru/about-tatneft/history-of-tatneft-group/recent-history--1990---2018/?lang=en.
Tatneft (2018c), 'Moody's povysilo kreditnyy reyting Tatnefti do Baa3, prognoz pozitivnyy', accessed 12 March 2019 at http://www.tatneft.ru/press-tsentr/press-relizi/more/5645/?lang=ru.
Vedomosti (2006a), 'S&P nakazalo Tatneft', accessed 12 March 2019 at https://www.vedomosti.ru/newspaper/articles/2006/08/28/sp-nakazalo-tatneft.
Vedomosti (2006b), 'Tatneft stala pervoi v Rossii', accessed 12 March 2019 at https://www.vedomosti.ru/newspaper/articles/2006/03/27/tatneft-stala-pervoj-v-rossii.
Vedomosti (2006c), 'Amerika proschaetsya s Tatneftiyu', accessed 12 March 2019 at https://www.vedomosti.ru/newspaper/articles/2006/11/30/amerika-proschaetsya-s-tatneftyu.
Vedomosti (2006d), 'Tatneft poluchila NPZ na Ukraine', accessed 12 March 2019 at https://www.vedomosti.ru/newspaper/articles/2006/10/30/tatneft-poluchila-npz-na-ukraine.
Vedomosti (2007), 'Berkut naletel na Tatneft', accessed 12 March 2019 at https://www.vedomosti.ru/newspaper/articles/2007/10/22/berkut-naletel-na-tatneft.
Vedomosti (2008a), 'Protsessing poshel', accessed 12 March 2019 at https://www.vedomosti.ru/newspaper/articles/2008/01/18/processing-poshel.
Vedomosti (2008b), 'Isk na $1.1 mlrd', accessed 12 March 2019 at https://www.vedomosti.ru/newspaper/articles/2008/06/07/isk-na-11-mlrd.
Vedomosti (2010), 'Snova na rynke', accessed 12 March 2019 at https://www.vedomosti.ru/newspaper/articles/2010/06/29/snova-na-rynke.
Vedomosti (2016), 'Tatneft, Lukoyl i Novatek voshli v top-10 mirovoy neftegazovoy otrasli', accessed 12 March 2019 at https://www.vedomosti.ru/business/articles/2016/10/25/662257-tatneft-lukoil-novatek.

Index